Table of Contents

Acknowledgements

A great many people contributed to the production of this publication. The *Forest Trees of Maine* Centennial Committee was instrumental in bringing this book into being. Members of the committee included Peter Lammert, Dan Jacobs, Merle Ring, Judy Tyler, Kevin Doran, Greg Miller, Jan Ames Santerre, Andy Shultz and Keith Kanoti. Special credit is due to Peter Lammert for updating the wood uses, Dan Jacobs for researching and writing the History section, and Greg Miller for producing the range maps. Others who contributed to the project include: Tom Whitworth, Scott Sawtelle, Tom Collins, Jen Wright, John Anastasio and Debbie Jacques.

Keith Kanoti revised the manuscript and acted as project manager.

The book was designed by David Deal and the manuscript was edited by Donna Stuart, both of Glen Group Inc.

Tom Rawinski, botanist with the US Forest Service Northern Research Station, reviewed the manuscript for technical accuracy.

Cover photo by Arthur Rogers, presented to Forest Commissioner Austin Wilkins by Governor Percival Baxter.

The glossary sketches were used with the generous permission of Dr. Michael A. Dirr.

The range maps are based on *Atlas of United States Trees* by the late Dr. Elbert L. Little.

The majority of the photographs were taken by the Maine Forest Service, Policy and Management Division field staff. Through bugwood.org, several photographers generously let us use images that we could not obtain ourselves. These include Paul Wray, Iowa State University (white oak acorns, swamp white oak acorns, northern red oak acorns, black oak acorns, butternut fruit, shagbark hickory fruit, bitternut hickory fruit, balsam fir cones, black willow leaves); The Dow Gardens Archives, Dow Gardens (mountain laurel flowers, rhododendron flowers, striped maple fruit); Chris Evans, River to River CWMA (bitternut hickory leaves); Bill Cook, Michigan State University Extension (beechnuts, American hornbeam fruit); and David J. Moorhead, University of Georgia (flowering dogwood flowers).

Other photography sources were obtained through USDA-NRCS PLANTS Database. These include Robert H. Mohlenbrock 1995, *Northeast wetland flora: Field office guide to plant species* (Atlantic white cedar cones); D.E. Herman, et al. 1996, *North Dakota tree handbook* (American basswood fruit, blue spruce cones/foliage, bur oak fruit, staghorn sumac fruit); and Steve Hurst (Sugar maple fruit, scarlet oak fruit, black ash fruit).

INTRODUCTION

In 1908 the Maine Forest Service released a booklet titled *Forest Trees of Maine*. In his 1910 Commissioner's report, Forest Commissioner Edgar Ring wrote of the popularity of the new publication: "For the *Forest Trees of Maine* there has been a large and constant demand which will very soon exhaust the edition. Possibly in order to meet the demands for this pamphlet it will be considered wise and money well spent to issue another edition." Now, 100 years later and in its 14th edition, *Forest Trees of Maine* remains the Maine Forest Service's most popular publication.

Since 1908, all editions of *Forest Trees of Maine* have had the same objective: to relate accurate information and to keep pace with new findings. As those who are familiar with *Forest Trees of Maine* will immediately notice, we have departed from the traditional format for this edition. This has allowed us to include color photographs, which have long been requested. For those who prefer the tried and true *Forest Trees of Maine* format, it will still be made available.

For the first time, range maps have been included. The maps are based on those of those of the legendary US Forest Service dendrologist, Dr. Elbert Little, who assisted with the 7th edition. The maps indicate the parts of the state where you are most likely to encounter each tree species. No map is perfect, and it is certainly possible to find a species outside of its indicated range.

The keys have been revised and, for the first time, a winter key has been included. To help you use the keys, sketches have been added to the glossary which illustrate many of the terms used. The keys are limited to the trees in the publication. For information on more complete keys, see Selected References on page 174.

The book contains information on 78 different tree species, including all of Maine's commercially important native tree species, as well as a few of the more common and important introduced trees. As with previous editions, no attempt has been made to include all the species in complicated groups, such as willows and hawthorns. When deciding which species to include in this edition, emphasis was placed on trees that occur in Maine's forests. With a few exceptions (e.g., horsechestnut, blue spruce, black walnut), species limited to ornamental plantings were excluded. Other introduced species were included if they have been commonly used in forest plantations (e.g., Norway spruce, Scots pine) or have escaped cultivation and are reproducing in forested areas (e.g., black locust, Norway maple). Several species are included that occasionally grow large enough to be considered small trees (e.g., bear oak, witch hazel, rhododendron, mountain laurel), but are more commonly found as shrubs.

Scientific names in this publication follow the Integrated Taxonomic Information System database: www.itis.gov.

Historic photographs found throughout the book are from the Maine Forest Service Archives and the Maine State Museum.

For more information about this publication or the Maine Forest Service, call 207-287-2791, e-mail us at: forestinfo@maine.gov or visit our website at: www.maineforestservice.gov

FOREWORD

I am privileged to be able to write a foreword for the centennial edition of *The Forest Trees of Maine*, this wonderful gift that the Maine Forest Service has provided for so long!

Suppose that someone invented a wonderful new machine. It can soak up the "greenhouse" gas carbon dioxide from the air and breathe out oxygen. It can pump huge amounts of water from the soil to reduce floods, while holding the soil together and helping to clean water that flows into streams and ponds. It can produce strong materials for building and fiber for paper. Pick up a stick of firewood. You hold the power of sunlight in your hands! This machine can store solar energy until we release it through fire. If we take care of the land where it grows, then our machine can be replaced by new machines just like it that will give us more of these wonderful things. Of course, the "machine" is not new at all—it is the tree— the tree that is beautiful in the forest and that forms beauty in the form of houses, furniture and the pages of a child's book.

Maine's forest trees are worth knowing. We don't have as many species as do some other states, but we have more acres of land covered with trees. Various sources estimate that we have nearly 96% as much forest as when Maine was first settled. More than in any other state, this great forest is privately owned, yet is more accessible to the public than is true in most areas.

Today too many children grow up in a world of television and computer games. They know little about where the food they eat, the clothing they wear, or the paper they write on comes from. Using this book to learn to identify trees, and to learn about trees and forests, can help to keep your children physically and mentally healthy and can be a great family activity. What fun it can be to explore the woods and see the diversity of trees! Can you identify one of the earliest to blossom in the spring—the serviceberry? Did you know that one shrub-like tree (witch hazel) does not blossom until the fall? Some—like the poplars (aspens), cherries, and white birch—are fast-growing "pioneers" that thrive in the full sun following a forest fire or timber harvest. Others, like sugar maple, can take root in deep shade and grow slowly for hundreds of years.

I hope that you will take the time to enjoy Maine's trees, and to use the wealth of information in this book to educate yourselves, your friends, and your children about this marvelous, renewable resource that is essential to Maine's quality of life.

—Dr. David Field, Professor Emeritus, School of Forest Resources, University of Maine

Logging crew, Upper Kennebec River, 1908

HISTORY SNAPSHOT 1908

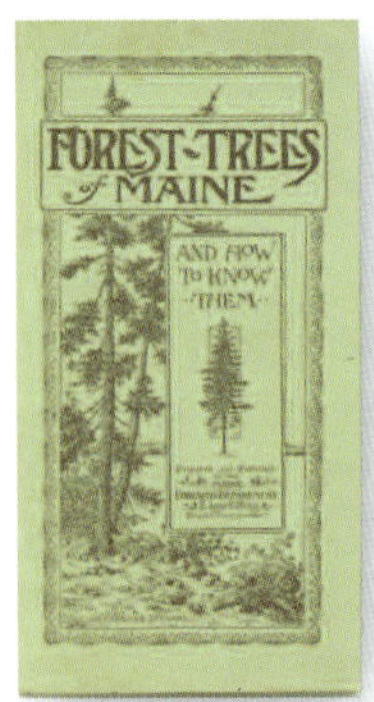

1908

The history of the most recognizable publication by the Maine Forest Service, *Forest Trees of Maine*, began in 1908.

While Mainers were enjoying the first edition of the *Forest Trees of Maine* that year, the nation as a whole was becoming increasingly hungry for forest resources. In the nation's capital, conservation issues and court rulings in Maine held the interest of the President.

In his 1908 State of the Union message, President Theodore Roosevelt declared, "Thanks to our own reckless use of our splendid forests, we have already crossed the verge of a timber famine in this country…"

In a speech at the White House that same year, the President applauded the State of Maine (particularly the Supreme Court of Maine) for an "exceedingly important judicial decision." The decision, which confirmed the legislature's authority to regulate timber harvesting, was viewed by President Roosevelt as a step towards "the wise utilization of forests….Such a policy will preserve soil, forests, waterpower as a heritage for the children and the children's children of the men and women of this generation."

In the conservation arena, 1908 was an important year: It marked the arrival of this great companion that has guided many people on journeys through the Maine woods.

Region II forest management and operations class, Dyer Brook 2008.

THEN AND NOW

A tremendous number of changes have occurred throughout Maine since the first printing of the *Forest Trees of Maine* in 1908. These changes, too numerable to count, have impacted all our lives, as well as our forest products industry. The table below illustrates some of the changes that have occurred in Maine over the past century.

THEN AND NOW	1908	2008
Forested Area (%)	75[1]	89
Population of Maine	694,466[2]	1,274,923
Population of U.S.	92,228,496	281,421,906
Harvest Volume (cords)	2,879,807	6,742,351*
Stumpage Price of Spruce (MBF**)	$5.49[1]	$135.00
Forestry Students at UMO	31[3]	50
Students at UMO	884	11,800±
Cost of Bangor Daily (year)	$6.00	$180.00±
Most Harvested Hardwood	Aspens[3]	Red Maple
Primary Use of Paper Birch	Spools[3]	Pulp
Gypsy Moth Infested Area of Maine	York and Cumberland	All Counties
Maine Indian Basketmakers	400	100
Primary Use of Black Ash Indian Baskets	Agricultural & Industrial	Collectables & Art

1 Kellogg, R.S. 1909. The Timber Supply of the United States, USDA Forest Service, Circular 166.
2 1909. Maine Register, State Year Book and Legislative Manual 1909-1910, No. 40, G.M. Donham, Portland, 1051pp.
3 Ring, E.E. 1910. Report of the Forest Commissioner, Maine, Kennebec Journal Print, Augusta, 110 pp.
* Harvest volume listed is for 2005. **MBF: Thousand board feet. Note: All values listed in nominal dollars.

1917

1925

1932

1951

A Few Precautions

The Maine woods are a pretty safe place; however, there are a few hazards anyone who is learning to identify trees should be aware of.

POISONOUS PLANTS

Maine has two species of poisonous plants you should learn to identify: poison ivy and poison sumac. Severe dermatitis can result when skin comes in contact with roots, stems, leaves, flowers, fruit or with implements or clothing that have come in contact with plant parts of either poison ivy or poison sumac. Smoke resulting from the burning of plant parts of either species is also poisonous.

POISON IVY, or mercury, is widely distributed throughout the state. It grows as an aerially-rooted climbing vine on trees or as a smooth, trailing vine or erect shrub along stonewalls, fencerows, roadsides and near water bodies.

The **leaves** are alternate, compound, with 3 very shiny, dark green leaflets. Leaflet margins are lobed, wavy, toothed or entire. The stalk of the terminal leaflet is much longer than those of the 2 lateral leaflets. Fall color is often a fiery red.

The **fruit** is a creamy-white, ribbed, globular, BB-sized drupe that occurs in axillary clusters.

POISON SUMAC is an uncommon species that is found throughout the southern part of the state and as far north as Penobscot County. It occurs as a small tree in low, wet swamps. It is particularly common around Mt. Agamenticus in southern Maine.

The **leaves** are alternate, 7–14 inches long, consisting of 7–13 leaflets along a smooth greenish-red rachis. Leaflets have entire margins, short stalks, are dark green and lustrous above with scarlet midribs, and paler and glabrous below. Twigs are without hairs.

Poison ivy (above) and poison sumac (below) are two plants everyone going into the woods should know how to identify and avoid.

The **fruit** is a globose, slightly compressed, thin-fleshed, ivory white or tawny white berry, about $\frac{1}{5}$ inch in diameter; it is borne in loose, pendent axillary clusters that ripen in September, but persist on the tree far into winter.

TICKS

About 13 different species of ticks live in Maine. One of these species, the deer tick (*Ixodes scapularis*) can transmit the bacterium that causes Lyme disease. Lyme disease frequently starts with a rash and flu-like symptoms, and if untreated may progress to neurological problems.

Ticks are most common in coastal and south-central Maine. When going into the woods in areas known to have high tick populations, it is wise to take some precautions to help avoid tick bites. For example:

- Tuck your pant legs into your socks and your shirt into your pants.
- Wear light-colored clothing so ticks can be seen more easily.
- Use a repellent containing DEET according the label directions. Pay special attention to treating shoes, socks and pant legs. Use caution in applying high-concentration products to the skin, especially on children.
- To protect pets, consult your veterinarian about tick repellents.
- Inspect yourself, your clothing, your children, your companion and your pets when you get in from the field. Ticks often attach to body folds, behind the ears and in the hair. If possible, shower and wash clothes immediately. Heat drying is effective in killing ticks.

Prompt removal of ticks from skin is very important. To remove a tick, grasp it as close to the skin as possible, preferably with tweezers, and pull gently but firmly until the tick lets go. Do not handle the tick with bare hands. Apply antiseptic to the bite. You should consult your physician if you develop a large rash at the site of the tick bite or if you remove an engorged deer tick.

For more information on ticks, visit the Maine Medical Center Research Institute Vector-borne Disease Laboratory's Lyme disease research Website: www.mmcri.org/lyme/lymehome.html.

Dog Tick

Male

Female

Deer Tick

Male

Female

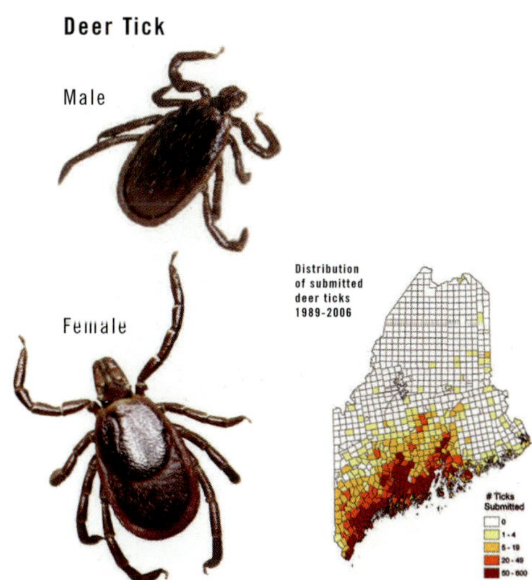

When supplies could not be brought to logging camps by water, they were transported over land on "tote" roads.

How to use the keys to identify a tree

This book contains keys to help you identify trees in both winter and summer. These are dichotomous keys; they work by giving you pairs of choices called couplets. Each couplet has the same number located on the left side of the key.

Begin at couplet number 1. Read both choices carefully to determine which matches the tree you are trying to identify. After you make your choice, the number at the right tells you which couplet to go to next. Go to that couplet and decide which choice matches your tree; repeat the process until you arrive at a name or species group for your tree. Turn to the page indicated and compare your tree to the species in the table to figure out the individual species.

The final step is to compare your tree to the pictures and drawings in the book. If they don't seem to match your specimen, don't be discouraged; return to the key and check to see if you made an error. Remember that leaves and bark can vary a lot even on the same tree, but the photograph can only show one example. To help you with the terms in the key, a glossary is provided on page 14.

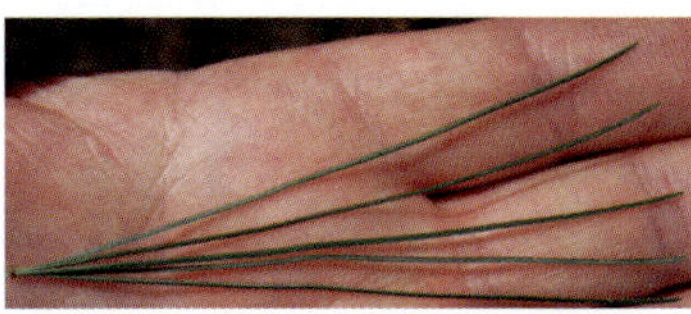

Example: We want to identify the tree these needles came from. Starting at the first couplet, choose the descriptions that fit the specimen. The lines in red indicate the correct choices in the key.

SEE GLOSSARY PAGE 14	GO TO
1. Leaves are needle-, awl- or scale-like; conifers	2
1. Leaves are broad and veined, not as above; hardwoods or broad-leaf trees	9
2. Leaves needle-like	3
2. Leaves awl- or scale-like, or both	7
3. Leaves flat, tips blunt, and occur singly	4
3. Leaves angular in cross section, tips pointed	5
4. Leaves taper, twigs limber; cones shorter than 1 inch	Eastern Hemlock p. 48
4. Leaves parallel-sided, twigs stiff; cones over 2 inches	Balsam Fir p. 46
5. Leaves occur singly, never clustered	Spruce p. 37
5. Leaves occur in clusters, also singly in larch	6
6. Leaves in clusters of 2–5 with papery sheath at base[1]	Pine p.25
6. Leaves in clusters[2] of 8 or more on spurs; papery sheath lacking	Tamarack p. 50

We now know the tree is a pine. We then go to the species table to figure out what species of pine it is.

Pines *The Important Distinctions*

	Eastern White Pine *Pinus strobus*	**Red Pine** *Pinus resinosa*	**Pitch Pine** *Pinus rigida*
NEEDLES			
NUMBER/ CLUSTER	5	2	3

The tree is Eastern white pine.

Winter Key

	SEE GLOSSARY PAGE 14	GO TO
1.	Leaves are evergreen	2
1.	Leaves are deciduous (dead leaves may remain attached)	10
2.	Leaves are needle- or scale-like; fruit is a cone; conifers	4
2.	Leaves are broad and flat, often curled in winter	3
3.	Leaves to 3 inches long	Mountain laurel p. 169
3.	Leaves 4–8 inches long	Rhododendron p. 170
4.	Leaves are needle-like	5
4.	Leaves are awl- or scale-like or both	8
5.	Leaves in clusters of 2, 3 or 5 with a papery sheath at the base (may be lacking in white pine)	Pine p. 25-36
5.	Leaves attached to the twig singly	6
6.	Leaves angular in cross section, will roll easily between the fingers	Spruce p. 37-45
6.	Leaves flat in cross section, will not roll easily between the fingers	7
7.	Leaves taper, attached to twig with tiny stem; bark thick, purple under scales	Hemlock p. 48
7.	Leaves parallel-sided, attached directly to twig with round base like tiny suction cup; bark thin with resin blisters	Fir p. 46
8.	Branchlets with awl-shaped leaves; leaves prickly	Juniper, Red cedar p. 56-58
8.	Branchlets with scale-like leaves; leaves not prickly	9
9.	Twigs flat; cones oblong, up to ½ inch; common statewide	Northern white-cedar p. 54
9.	Twigs slightly flattened; cones ¼ inch, rounded; rare tree of southern and midcoast Maine	Atlantic white cedar p. 52
10.	Older twigs with many short spur branches less than ¼ inch long; fruit a cone	Tamarack p. 50
10.	Spur branches lacking or if present are longer than ¼ inch; fruit not a cone	11
11.	Leaf scars are opposite	12
11.	Leaf scars are alternate	17
12.	3 bundle scars (may be obscured in flowering dogwood, see 15)	13
12.	More than 3 bundle scars	16
13.	More than 2 bud scales	Maple p. 70-85
13.	2 bud scales	14
14.	Buds long and narrow, base of terminal bud swollen; bud scales covered with minute, scale-like particles	Nannyberry p. 166
14.	Buds not long and narrow; bud scales not covered with minute, scale-like particles	15
15.	Lateral buds hidden; rare small tree of southwestern Maine	Flowering dogwood p.171
15.	Lateral buds conspicuous; common small trees	Maple p. 70-85
16.	Buds shiny and sticky in spring before flowering	Horsechestnut p. 158
16.	Buds not shiny; bud scales covered with minute, scale-like particles	Ash p. 127-133
17.	Twigs armed with spines, thorns or branches ending in a spine	18
17.	Twigs are unarmed	20
18.	Armed with paired spines less than 1 inch long	Black locust p. 162
18.	Armed with thorns or branches ending in a spine greater than 1 inch long	19
19.	Armed with thorns that occur just above the leaf scar	Hawthorn p. 147
19.	Armed with branches ending in a spine	Canada plum p. 146
20.	Buds not visible	Black locust/Honey locust p. 162-164
20.	Buds visible	21
21.	Leaf scar nearly encircling the bud	22
21.	Leaf scars extending less than ¾ of the way around the buds	23
22.	Buds covered by scales; twigs not hairy; bark mottled in color	Sycamore p. 159
22.	Buds naked; twigs very hairy; bark with prominent lenticels	Staghorn sumac p. 168
23.	Pith chambered or diaphragmed	24
23.	Pith solid	25
24.	Pith chambered	Butternut or Walnut p. 155-157
24.	Pith diaphragmed; uncommon tree of swamps in southern Maine.	Black gum p. 160

25.	Buds naked	26
25.	Buds covered by one or more scales	27
26.	Terminal buds scalpel-shaped	**Witch-hazel p. 167**
26.	**CAUTION POISONOUS** Terminal buds ovoid	**Poison sumac p. 6**
27.	Buds covered with a single cap-like scale	**Willow p. 68**
27.	Buds covered by 2 or more scales	28
28.	A single bundle scar; crushed twigs aromatic; rare tree of southwest Maine	**Sassafras p. 165**
28.	More than 1 bundle scar	29
29.	Catkins present	30
29.	Catkins absent	32
30.	Buds stalked; 2 types of catkins present	**Alder p. 102**
30.	Buds sessile; 1 type of catkin present	31
31.	Bud scales with tiny grooves; gray bark with loose vertical scales	**Eastern Hop-Hornbeam p. 98**
31.	Bud scales without grooves; bark peeling or blocky, not with loose vertical scales	**Birch p. 86-96**
32.	Terminal bud absent, the end bud is pseudo-terminal (except on spur shoots)	33
32.	Terminal bud present	38
33.	Up to 4 bud scales (except on spur shoots)	34
33.	5 or more bud scales	36
34.	Bud scales deep red; fruit hard and round, borne in cymes attached to a bract	**American basswood p. 134**
34.	Bud scales other than deep red; fruits not attached to a bract	35
35.	Pith star-shaped (rare tree)	**American chestnut p. 124**
35.	Pith round, fruit borne in catkin-like cones	**Birch p. 86-96**
36.	Buds four-angled, square in cross section; stem fluted, gun metal gray	**American Hornbeam p. 100**
36.	Buds not four-angled, trunk and bark not as above	37
37.	Buds with small vertical grooves, yellow-green, round in cross section; grayish-brown bark peeling into vertical scales **Eastern Hop-Hornbeam p. 98**	
37.	Buds without small vertical grooves, brown, often laterally flattened; bark corky, ridged, often with alternating light and dark layers in cross section **Elm p. 136-137**	
38.	Leaf scars very long and narrow, several times longer than broad	39
38.	Leaf scars broader, at most 3 times longer than broad.	40
39.	Second bud scale more than ½ the length of the bud; buds always uniformly colored	**Mountain ash p. 150**
39.	Second bud scale less than ½ the length of the bud; buds sometimes bi-colored—reddish and greenish	**Serviceberry p. 148**
40.	Pith star-shaped	41
40.	Pith round or if angled without 5 points.	45
41.	Buds clustered toward the tip of the twig	**Oak p. 106-123**
41.	Buds not clustered toward the tip of the twig	42
42.	Terminal bud more than ⅜ inch long; bud scales loose; older trees with gray shaggy bark	**Shagbark hickory p. 152**
42.	Terminal buds less than ⅜ inch long; bud scales not loose; bark of older trees fissured and ridged or smooth, not shaggy	43
43.	Buds sulfur yellow (rare tree of southwestern Maine)	**Bitternut hickory p. 154**
43.	Buds brown to reddish-brown	44
44.	Lowest bud scale centered over the leaf scar; wood diffuse porous	**Aspen/Poplar p. 61-67**
44.	Lowest bud scale not centered over the leaf scar; wood ring porous (rare tree)	**American chestnut p. 124**
45.	Buds long and narrow, several times longer than broad, diverge from the twig at a wide angle; bark smooth gray or often pockmarked with small cankers **American beech p. 104**	
45.	Buds not long and narrow, do not diverge from twig at wide angles.	46
46.	Nodes often clustered toward the ends of twig; bark of dead branchlets yellowish-orange	**Alternate-leaved dogwood p. 172**
46.	Nodes not clustered; dead branchlets not yellowish-orange	47
47.	Buds stalked; pith triangular	**Alder p. 102**
47.	Buds sessile; pith round	**Cherry p. 139-145**

SUMMER KEY

No.	Description	GO TO
1.	Leaves are needle-, awl- or scale-like; conifers	2
1.	Leaves are broad and veined, not as above; hardwoods or broad-leaf trees	9
2.	Leaves needle-like	3
2.	Leaves awl- or scale-like, or both	7
3.	Leaves flat, tips blunt, and occur singly	4
3.	Leaves angular in cross section, tips pointed	5
4.	Leaves taper, twigs limber; cones shorter than 1 inch	Eastern Hemlock p. 48
4.	Leaves parallel-sided, twigs stiff; cones over 2 inches and upright	Balsam Fir p. 46
5.	Leaves occur singly, never clustered	Spruce p. 37-45
5.	Leaves occur in clusters, also singly in larch	6
6.	Leaves in clusters of 2–5 with papery sheath at base[1]	Pine p. 25-36
6.	Leaves in clusters[2] of 8 or more on spurs; papery sheath lacking	Tamarack p. 50
7.	Branchlets with prickly, awl-shaped leaves; cones are berry-like	Juniper/Eastern Redcedar p. 56-58
7.	Branchlets with scale-like leaves; leaves not prickly; cones un-berry-like	8
8.	Twigs flat; cones oblong, woody, up to ½ inch; wood slightly aromatic	Northern White Cedar p.54
8.	Twigs slightly flattened; cones ¼ inch, rounded, leathery; wood strongly aromatic	Atlantic White Cedar p. 52
9.	Leaves opposite, trees only	10
9.	Leaves alternate	15
10.	Leaves simple	11
10.	Leaves compound	13
11.	Leaf margin serrate	Nannyberry p. 166
11.	Leaf margin lobed or entire	12
12.	Leaf margin lobed	Maple p. 70-85
12.	Leaf margin entire	Flowering Dogwood p. 171
13.	Leaves palmate	Horsechestnut p. 158
13.	Leaves pinnate	14
14.	3–5 leaflets, lobed, coarse teeth	Boxelder p. 84
14.	5–13 leaflets	Ash p. 127-133
15.	Leaves simple	16
15.	Leaves compound	36
16.	Leaf margin entire, wavy, or lobed	17
16.	Leaf margin toothed or serrate	24
17.	Leaf margin entire	18
17.	Leaf margin wavy or lobed	21
18.	Leaves thin, veins parallel	Alternate Leaf Dogwood p. 172
18.	Leaves thick and leathery, net-veined	19
19.	Pith diaphragmed; leaves 2–5 inches long	Black Tupelo p. 160
19.	Pith not diaphragmed	20
20.	Leaves to 3 inches long	Mountain Laurel p. 169
20.	Leaves 4–8 inches long	Roseberry Rhododendron p. 170
21.	Leaf margin wavy toward tip; base of leaf one-sided	Witch-Hazel p. 167
21.	Leaf margin lobed or wavy throughout	22
22.	Leaf petiole hollow and covers bud; numerous main leaf veins radiate from base	American Sycamore p. 159
22.	Leaf petiole neither swollen nor hollow; leaves with one main vein	23

[1] Papery sheath on white pine drops in late August
[2] Larch leaves are borne singly on elongating shoots

23.	Twigs angular; pith star-shaped	**Oak p. 106-123**
23.	Twigs round, spicy odor and taste; leaves 0–3 lobed	**Sassafras p.165**
24.	Leaf margin singly toothed or serrated	25
24.	Leaf margin doubly serrated	31
25.	Teeth hooked, prominent; fruit a bur	26
25.	Teeth not hooked, fruit not a bur	27
26.	Pith star-shaped; buds blunt; bark brown	**American Chestnut p. 124**
26.	Pith round; buds long, pointed; bark gray	**American Beech p. 104**
27.	Leaf base one-sided, leaf cordate; pith not symmetrical	**American Basswood p. 134**
27.	Leaf base even; pith symmetric in cross section	28
28.	Leaves long and narrow; petioles short without glands; buds with a single, cap-like scale	**Willows p. 68**
28.	Leaves broad, or if narrow with glands on petiole; buds with several scales	29
29.	Leaf petiole usually long, flat, except rounded in balsam poplar; pith star-shaped	**Aspen/Poplar p. 61-67**
29.	Leaf petiole short, not flat; pith round	30
30.	Twigs pungent when broken; glands on petiole	**Cherry, Plum p. 139-145**
30.	Twigs odorless; leaf petiole glandless; buds slender, twisted at tip, silky within	**Serviceberry p. 148**
31.	Leaf base one-sided, surface sand-papery	**Elm p. 136**
31.	Leaf base even, surface smooth	32
32.	Branches with thorns 1 inch or more long	**Hawthorn p. 147**
32.	Branches without thorns	33
33.	Pith triangular; buds stalked, smooth	**Speckled Alder p. 102**
33.	Pith not triangular; bud scales overlapping	34
34.	Leaves hairy on both surfaces; pith green	**Eastern Hop-Hornbeam p. 98**
34.	Leaves if hairy only so on one surface; bark smooth	35
35.	Stem fluted; bark smooth, gun-metal gray	**American Hornbeam p. 100**
35.	Stem not fluted; bark white, yellow, or red to dark brown	**Birch p. 87-96**
36.	Leaflets with margin entire	37
36.	Leaflets with serrated margin	38
37.	Twigs with paired spines; 7–19 leaflets	**Black Locust p. 162**
37.	Twigs spineless; 7–13 leaflets; leaflets poisonous	**Poison-Sumac** ** p. 6**
38.	Leaflets ½ inch long with fine, rounded teeth	**Honeylocust p. 164**
38.	Leaflets over 1 inch long	39
39.	Pith chambered or diaphragmed	**Black Walnut/Butternut p. 155–157**
39.	Pith solid	40
40.	5–7 leaflets; pith star-shaped	**Shagbark/Bitternut Hickory p. 152–154**
40.	11–31 leaflets	41
41.	Twigs smooth; 11–17 leaflets; buds large	**Mountain Ash p. 150**
41.	Twigs densely hairy; 11–31 leaflets; buds small	**Staghorn Sumac p. 168**

** See also Poison ivy, page 6

GLOSSARY

Structure in brackets indicates part to which the term applies.

Abortive *[fruit]* Not developed completely.

Alternate *[arrangement of leaves or buds]* Not opposite on sides of twig.

Appressed Pressed close or lying flat against something.

Awl-shaped *[leaf]* Narrow and tapering to a sharp point.

Axillary Growing from the Axil. The angle between the upper side of a leaf or stem and the supporting stem or branch.

Basal disc *[fruit]* A plate-like structure on the base of a fruit.

Bloom A whitish covering; usually on new shoot growth or fruit.

Bole The main stem of a tree; usually the part that is commercially useful for lumber or other wood products.

Bract A leaf-like structure which is attached to a flower, a fruit or to its stalk.

Branchlet Shoot growth of the latest growing season.

Broadleafed Having relatively broad rather than needle-like or scale-like leaves.

Bur *[fruit]* A prickly or spiny husk enclosing the seed.

Capsule *[fruit]* A dry fruit enclosing more than one seed and splitting freely at maturity.

Catkin A compact, cylindrical cluster of flowers of the same sex.

Chambered *[pith]* With hollow cavities separated by discs or plates.

Compound *[leaf]* A leaf composed of smaller leaf units or leaflets.

Conical Wide at the base and gradually tapering to a point; circular in cross section.

Conifer Cone-bearing trees; the "evergreens."

Cordate *[leaf]* Heart-shaped at the petiole end or base.

Corymb A flat-topped floral cluster with outer flowers opening first.

Cup *[fruit]* The scaled, concave basal portion of oak fruit.

Cyme A flattened flowering structure, center flowers bloom earliest.

Deciduous *[leaves]* All leaves drop in the autumn; not evergreen.

Diaphragmed *[pith]* Solid but divided into sections by firmer discs.

Drupe *[fruit]* Fleshy outside, hard and stone-like inside.

Ellipsoid Tapers equally at both ends; more than twice as long as broad.

Elliptical Like an ellipse; flat and tapering equally at both ends.

Entire *[leaf]* Margin of leaf without teeth, lobes, or divisions.

Fascicle *[leaf]* A cluster of conifer leaves.

Fluted *[stem]* With alternating, rounded depressions and ridges.

Fruit The seed-bearing part of a tree.

Glabrous Smooth, without hairs

Glands Generally raised structures at the tips of hairs, or on a leaf, petiole, or twig.

Globose Spherical or globe-shaped.

Habitat The place where a plant usually grows, e.g. rocky, moist, well-drained, etc.

Hardwood Term used to describe all broad-leaved trees. These tree species are typically deciduous, retaining their leaves only one growing season. Despite the term, some "hardwoods," such as the aspens, have wood that is relatively soft.

Head A compact aggregate of flowers or fruit on a common stalk.

Husk *[fruit]* The somewhat leathery, outer covering of a fruit sometimes capable of splitting along well-defined lines.

Invasive Not native to and tending to spread widely in a habitat or environment, sometimes displacing native species.

Lance-shaped Long and tapering; several times longer than broad; broadest at the base.

Leaf Stalk (petiole) and blade of hardwoods; needles and scales of conifers.

Leaflets Smaller leaf units which together form a compound leaf.

Lenticel *[bark]* Corky, raised pores on woody parts with openings for air-gas exchange.

Linear *[leaf]* Much longer than broad with parallel margins.

Lobed *[leaf]* With large, rounded or pointed projections along the leaf margin. Projection formed by indentations of the leaf margin.

Margin *[leaf]* The edge, perimeter, or portion forming the outline.

Midrib *[leaf]* The large central vein.

Oblong Longer than wide with nearly parallel sides.

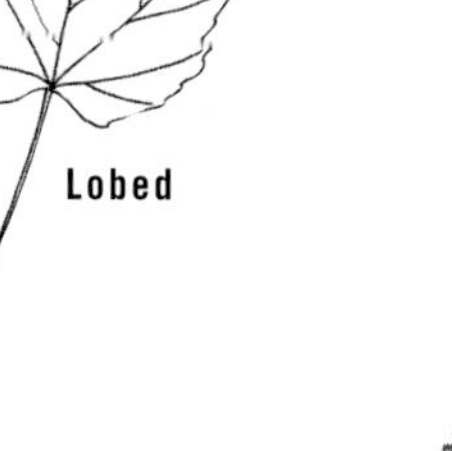

Obovate Egg-shaped in outline; broadest above the middle.

Opposite *[arrangement of leaves or buds]* Directly across from one another on a common axis, or twig.

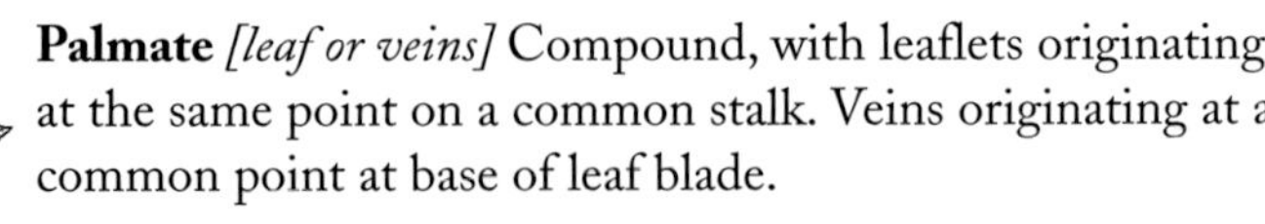

Oval Somewhat elliptical; less than twice as long as broad.

Ovate Egg-shaped in outline; broadest below the middle.

Ovoid An egg-shaped solid.

Palmate *[leaf or veins]* Compound, with leaflets originating at the same point on a common stalk. Veins originating at a common point at base of leaf blade.

Panicle A loosely branched, pyramidal cluster of flowers.

Pendulous Drooping or hanging downward.

Petiole *[leaf]* The stalk that supports the leaf blade.

Pinnate *[leaf or vein]* Compound, with leaflets along a common rachis or stalk. Veins originating along a common mid-vein.

Pistillate Containing female portions of flowers, or the pistils.

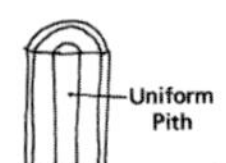
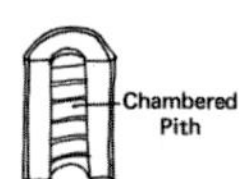
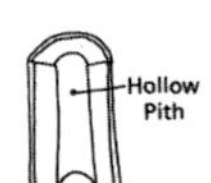

Pith The central, soft part of the stem.

Prickle A small spine-like growth.

Pseudo-terminal *[bud]* When the bud on the end of a twig has a leaf scar located directly below.

Pubescent Covered with hairs.

Raceme Numerous stalked flowers or fruit along a common axis.

Rachis The common stalk in a compound leaf to which the leaflets are attached.

Ranked *[leaves]* Arranged in rows or files.

Samara A winged fruit, e.g. ash, maple.

Scales *[bud]* Small, modified leaves on the outer surface of buds.

Scales *[cone]* The basic structures that enclose the seeds.

Scale-like *[leaf]* Small, generally overlapping, triangular-shaped leaves of some conifers.

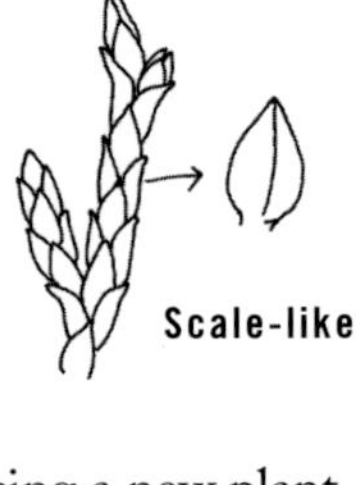

Seed That part of the fruit capable of germinating and producing a new plant.

Serrate *[leaf]* Margins with a saw-tooth outline. Doubly serrate: with small teeth on the larger teeth.

Sessile Attached directly by the base without an intervening stalk.

Shrub A woody, many-stemmed plant, usually under 15 feet in height at maturity, which branches from its base.

Simple *[leaf]* A single leaf composed of a single blade. Not compound.

Smooth Without hairs, glands, or any roughness.

Softwood Term used to describe all needle-leaved trees. These species are typically evergreen, retaining their leaves through two or more growing seasons. Larches, including tamarack, are exceptions, being deciduous "softwoods."

Solid *[pith]* Without cavities or sections separated by discs.

Spike A flower stalk.

Spinescent Having a spine or spines; or terminating in a spine.

Spur A short, extremely slow-growing, woody twig projection.

Staminate Containing male portions of flowers, or the stamens.

Stipule A tiny, leafy, sometimes spiny projection arising at the base of a petiole.

Stomate *Plural stomata* Small pore on a leaf used for gas exchange.

Stone The "bony" or stony pit of drupes.

Style The usually slender part of a pistil, situated between the ovary and the stigma.

Toothed *[leaf]* With moderate projections along the margin.

Tree A woody plant, generally single-stemmed, that reaches a height of more than 15 feet at maturity and a diameter of 3 inches or more measured at 4½ feet above the ground.

Umbel A group of flowers or fruit whose stalks have a common point of attachment.

Unequal *[leaf base]* Base parts of blade on either side of midrib are uneven.

Valve-like *[bud scales]* Meet at their margins and do not overlap.

Wavy *[leaf margin]* Undulating but smooth; not toothed nor lobed.

Whorl *[leaves or branches]* More than two originating at the same level on a common axis.

LEAF SHAPES

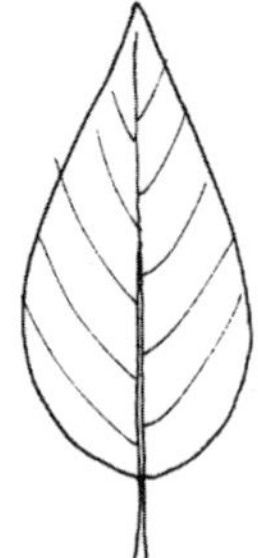

Ovate

Lanceolate
(Lance-shaped)

Cordate
(Heart-shaped)

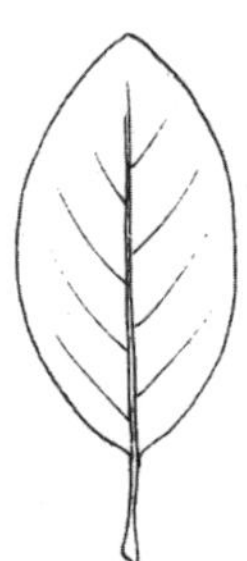

Elliptical

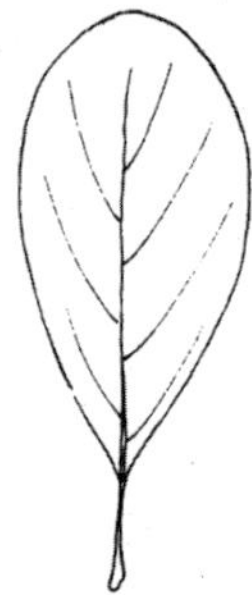

Obovate

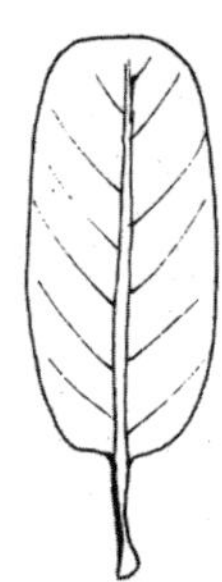

Oblong

Linear

MARGINS

Entire

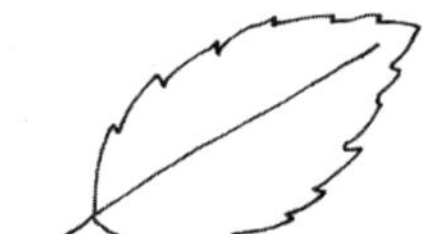

Serrate
(Toothed)

Crenate
(Round toothed)

Doubly-Serrate
(Doubly toothed)

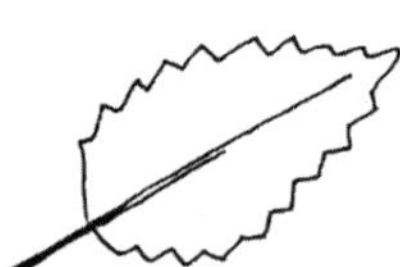

Dentate
(Coarsely toothed)

LEAF STRUCTURES

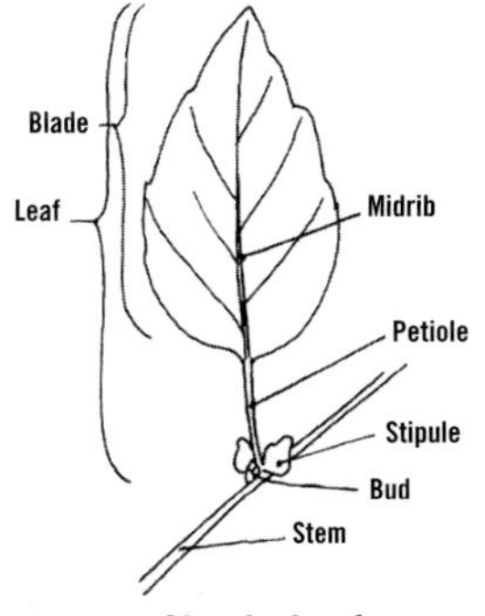

Simple Leaf

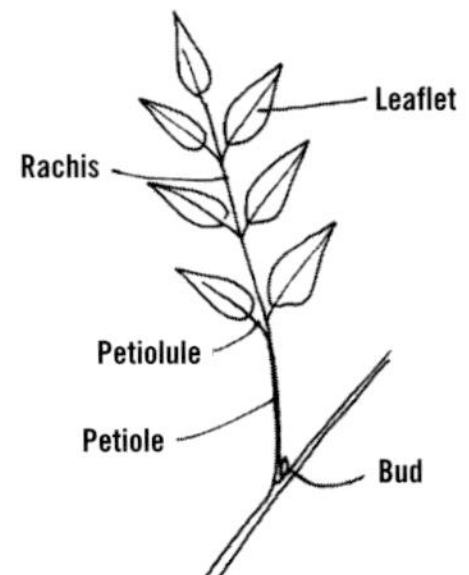

Pinnately Compound Leaf

Palmately Compound Leaf

NEEDLE TYPES

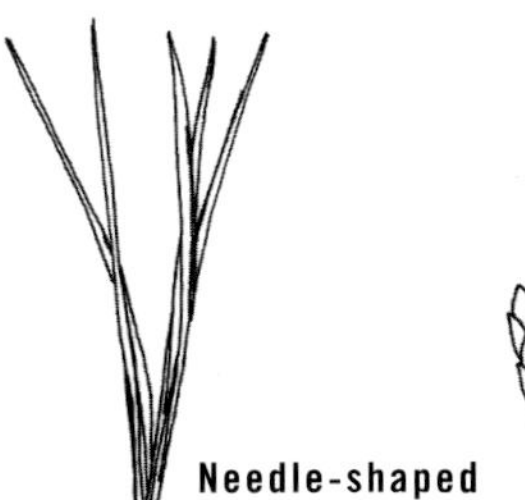

Needle-shaped

Scale-shaped

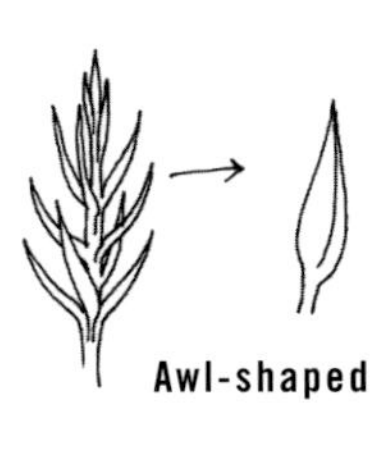
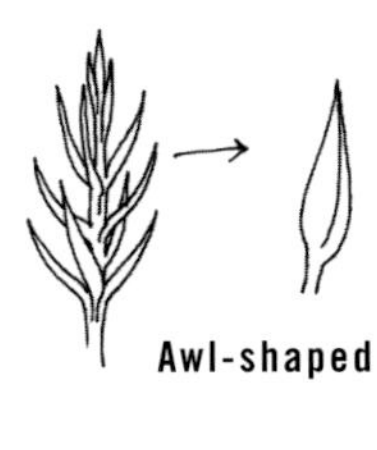

Awl-shaped

TWIG STRUCTURE

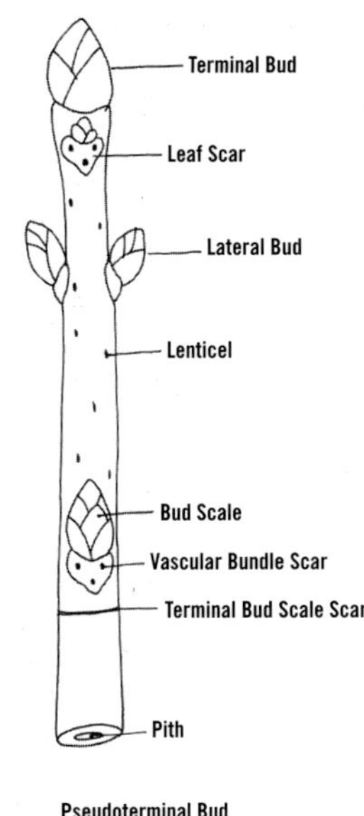

FLOWER TYPES

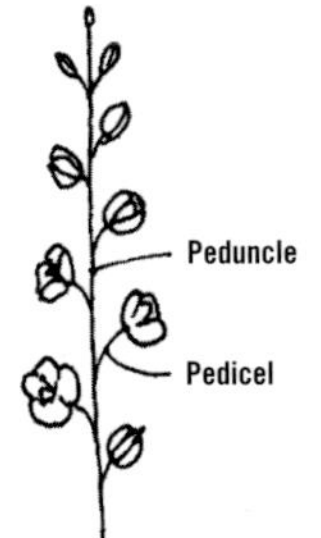

Raceme

Spike

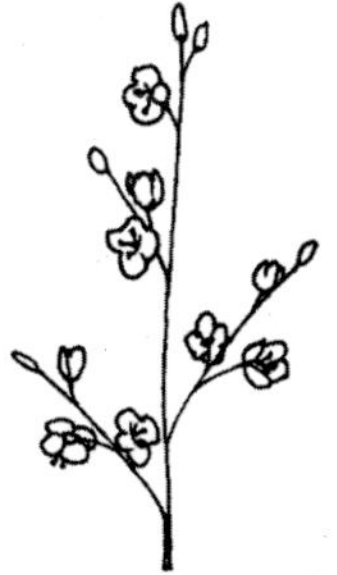

Panicle (of Racemes)

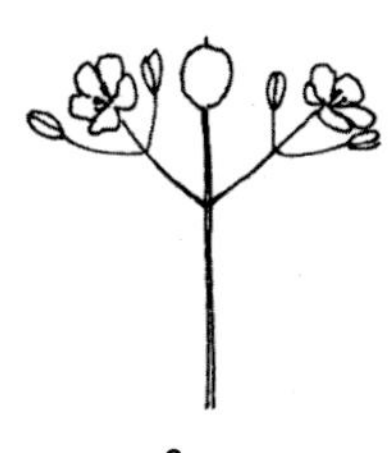

Cyme

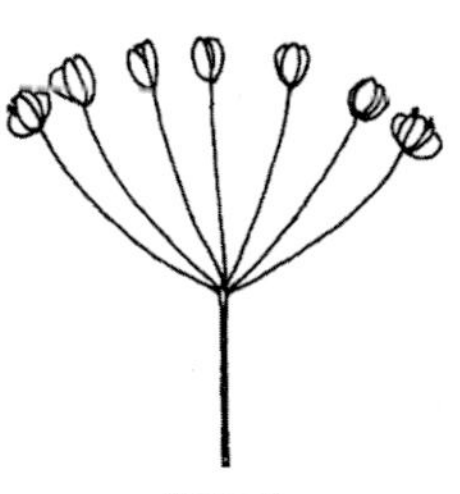

Umbel

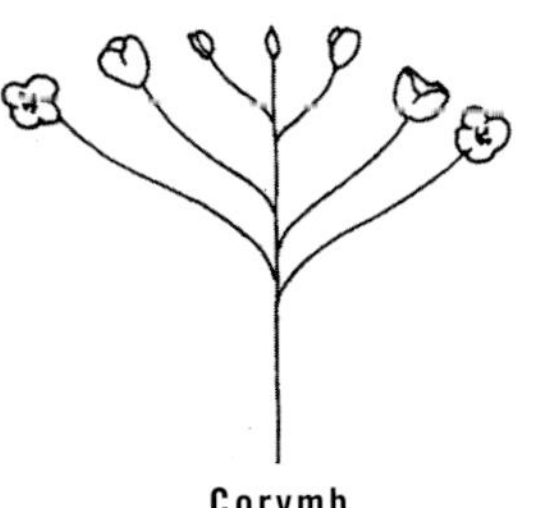

Corymb

TREE PARTS AND FUNCTIONS

A tree has three major parts: **roots, trunk** and **crown**.

Large roots anchor the tree and store foods which are manufactured in the leaves. **Small roots** and **root hairs** absorb water and dissolved mineral salts from the soil. These raw materials are conducted upward to the leaves where they are utilized in the synthesis of necessary plant food. Air must be present in the soil for the roots to live, although some species can endure several months of flooding.

The **trunk** is the main body of the tree. In the center of the trunk is the pith. Next to the pith is the **heartwood** which is composed of dead cells and serves as support. On the outer side of the heartwood is the **sapwood**, which contains the sap conducting tubes. Sapwood is usually lighter in color, but it darkens with age and becomes heartwood. Heartwood and sapwood together comprise the **xylem**. Outside the sapwood is the **cambium**, a thin layer of cells, which annually produces new sapwood inwardly and new **inner bark** outwardly. The cambium produces diameter growth, and callus growth around open wounds. The **inner bark** or **phloem** is outside the cambium and carries food from the leaves downward to nourish the cambium and growing parts. The **outer bark** is the outer-most part of the tree. Essentially, it is composed of dead cork cells and protects the inner bark from mechanical injury, drying or disease; it also insulates the phloem from extremes of heat and cold. Damage to the phloem causes interference with food movement to growing parts below the injury. Girdling of a tree through its inner bark will kill the tree. **Wood or medullary rays** radiate out from the center, and serve in lateral conduction and as food storage areas. They are most conspicuous in a cross-sectional view.

The **crown** is composed of branches, twigs, buds, leaves, flowers and fruit. The process of **photosynthesis** occurs in the **leaves**. Using energy produced by sunlight, the leaves combine carbon dioxide from the air and water from the soil to produce **carbohydrates**. Oxygen is released in the process. Carbohydrates plus fats and proteins are the plant foods necessary for growth and respiration of the tree. **Flowers** and **fruit** are important in reproduction.

ANNUAL RINGS

The **yearly growth** of a tree can be compared to the annual placement of hollow wooden cones, one on top of the other. Each cone would represent a single year's growth over the entire stem. At the beginning of each new growth period, new wood cells are large and thin-walled, and form the **springwood** or **early wood**. As the growing season progresses, the smaller, thicker-walled cells of the **summerwood** or **late wood** are produced. The darker appearance of the late wood delineates the **annual ring** of growth put on by a tree. The age of a tree, at any desired point along the trunk, can be determined by counting these annual rings.

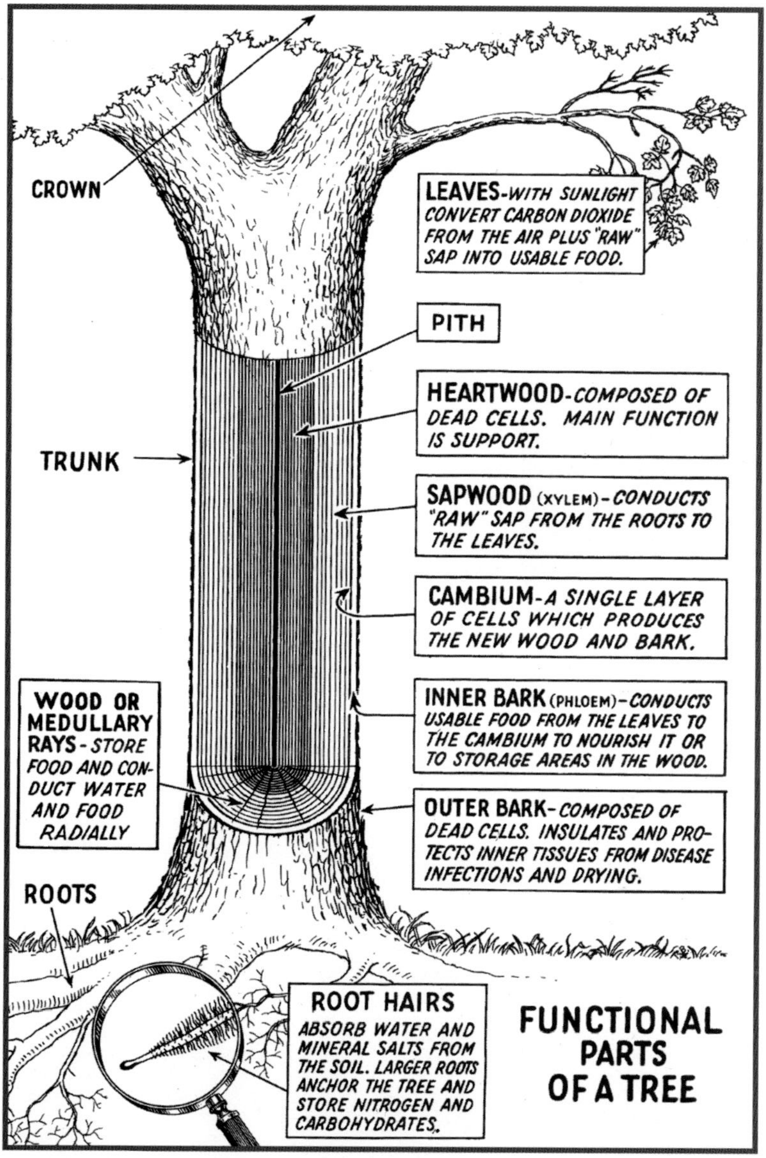

CROWN

LEAVES—WITH SUNLIGHT CONVERT CARBON DIOXIDE FROM THE AIR PLUS "RAW" SAP INTO USABLE FOOD.

PITH

HEARTWOOD—COMPOSED OF DEAD CELLS. MAIN FUNCTION IS SUPPORT.

TRUNK

SAPWOOD (XYLEM)—CONDUCTS "RAW" SAP FROM THE ROOTS TO THE LEAVES.

CAMBIUM—A SINGLE LAYER OF CELLS WHICH PRODUCES THE NEW WOOD AND BARK.

INNER BARK (PHLOEM)—CONDUCTS USABLE FOOD FROM THE LEAVES TO THE CAMBIUM TO NOURISH IT OR TO STORAGE AREAS IN THE WOOD.

WOOD OR MEDULLARY RAYS—STORE FOOD AND CONDUCT WATER AND FOOD RADIALLY

OUTER BARK—COMPOSED OF DEAD CELLS. INSULATES AND PROTECTS INNER TISSUES FROM DISEASE INFECTIONS AND DRYING.

ROOTS

ROOT HAIRS ABSORB WATER AND MINERAL SALTS FROM THE SOIL. LARGER ROOTS ANCHOR THE TREE AND STORE NITROGEN AND CARBOHYDRATES.

FUNCTIONAL PARTS OF A TREE

The relationship between tree age and size can be very deceptive. Many tree species can survive in a shaded understory for years, with nearly microscopic growth rings. These tree species are called shade tolerant. When the canopy is partially or wholly removed by harvest, mortality or weather, increased available sunlight allows accelerated growth. This process is known as release. Some long-lived species, such as red spruce, can survive for decades in the understory, then for centuries in the overstory. Others are shade tolerant but short lived; for example, balsam fir rarely exceeds 100 years in total age. Still other species such as quaking aspen and paper birch are intolerant of shade and relatively short-lived. An observer with knowledge of tree species and growth characteristics can deduce the history of a forest without cutting or boring holes in stems to count rings.

Autumn Coloration

Autumn foliage coloration, one of Maine's greatest aesthetic assets, is enjoyed every year, generally with little appreciation for the processes responsible for it. Most hardwoods produce dramatic leaf coloration if climactic conditions are favorable, while conifers usually produce only weak coloration of yellow and brown.

Hardwoods contain green, yellow and orange pigments in their leaves. Chlorophyll allows the green to be the most prominent of the pigments; however, the green pigment is also the least stable. It is repeatedly produced and destroyed throughout the summer and masks the carotenoid pigments (xanthophyll and carotene) which give the yellow and orange shades. As autumn approaches, chlorophyll is destroyed faster than it is produced. As the chlorophyll disappears, the carotenoid pigments begin to show. The purple and brilliant red shades become visible from the production of anthocyanin pigments, which are also capable of masking the carotenoids. Tannins cause brown shades in some species.

Certain conditions favor maximum autumn coloration. They include adequate summer rainfall, adequate sugar accumulations in the leaves and prolonged periods of cool, bright, sunny weather without severe frosts. Frost is not an essential element for leaf coloration. In fact, weakened trees occasionally color in mid-summer.

Although variations are numerous, a general guideline to autumn tree coloration is listed below.	
YELLOWS	tamarack, green ash, black ash, basswood, beech, birch, butternut, elm, boxelder, mountain maple, silver maple, striped maple, sugar maple, mountain-ash, poplar, serviceberry, willow, witch-hazel.
RED/SCARLET	hornbeam, red maple, mountain maple, sugar maple, black oak, red oak, scarlet oak, white oak, sumac, tupelo.
ORANGE	sugar maple.
BROWNS	black oak, beech.
PURPLE	white ash.

Although much less appreciated than fall color, the subtle color of Maine's trees in early spring can be just as dramatic.

CONIFERS

Often referred to as "softwoods", conifers belong to the group of plants known as the gymnosperms. Conifers are cone bearing trees and shrubs that have needle or scale like leaves and resinous wood. All of Maine's conifers except for tamarack are evergreen. There are 16 species of conifer native to Maine, 13 of these are trees and three others are more commonly found as shrubs. Several other species of conifer native to other parts of the world are commonly planted in Maine for both ornamental and timber production purposes.

Photo location: Penobscot Experimental Forest, Bradley, Maine.

PINES *The Important Distinctions*

	Eastern White Pine *Pinus strobus*	**Red Pine** *Pinus resinosa*	**Pitch Pine** *Pinus rigida*
NEEDLES			
NUMBER/ CLUSTER	5	2	3
DESCRIPTION	Slender, flexible, 3–5 inches	Straight, flexible, 4–6 inches	Stout, not flexible, usually twisted, grow at right angles to the branchlets, 3–5 inches
COLOR	Bluish-green	Dark green	Dark yellow-green
SHEATH	Shed in late August	Persists	Persists
CONES			
LENGTH	4–8 inches	1½–2¼ inches	1½–3½ inches
DESCRIPTION	Borne on a long stalk; thin smooth scales without prickles	Borne on short stalks; scales without prickles. Several basal scales remain on branches when cone drops.	Borne on a short stalk, having prickles on the cone scales, flat-based when completely open. Often remain on branches for 10–12 years.

	Jack Pine *Pinus banksiana*	**Scots Pine** *Pinus sylvestris*
NEEDLES		
NUMBER/ CLUSTER	2	2
DESCRIPTION	Stout, flat, twisted, ¾–1½ inches	Stout, stiff, twisted, 1½–3 inches
COLOR	Light yellow-green, later becoming dark green	Dull blue-green
SHEATH	Persists	Persists
CONES		
LENGTH	1½–2 inches	1–2 inches
DESCRIPTION	Much curved inward, without stalk. Prickles minute. Often remain on branches for many years.	Egg shape, borne on a short stalk, scales with occasional prickles.

EASTERN WHITE PINE *Pinus Strobus* L.

Eastern white pine has been an important tree for the people of what is now the State of Maine for hundreds, if not thousands, of years. Therefore, it is no coincidence that Maine has come to be known as the "Pine Tree State." Recognizing its importance, in 1895 the Maine legislature designated the "Pine Cone and Tassel" as Maine's official floral emblem. In 1945 the legislature Resolved: "That the white pine tree be, and hereby is, designated the official tree of the State of Maine."

The availability and high quality of white pine lumber has played an important part in the development and economy of Maine since 1605, when Captain George Weymouth of the British Royal Navy collected samples here and brought them back to England for display. The shortage of ship masts in Europe led to England's Broad Arrow Policy in 1691, whereby pines 24 inches or more in diameter within 3 miles of water were blazed with **the mark of the**

Maine is known as the "Pine Tree State" and the Eastern white pine is the official tree of the State of Maine.

broad arrow; such trees to be reserved for use in the Royal Navy. The term **King's Arrow Pine** originated from this policy. Most of the accessible virgin pine was cut by 1850. Lumber production reached its peak in 1909, but white pine is still a valuable species that contributes greatly to the economy of the state.

White pine occurs in all localities in the state in moist situations, on uplands and on sandy soil, but develops best on fertile, well-drained soils. On sandy soil it often becomes established in pure or nearly pure stands. It is one of the major species planted in the state. The tree grows rapidly both in height and diameter, making an average growth in height of 1 foot or more each year.

When growing in the open, the young tree is symmetrical and conical in outline except when deformed by white pine weevil. White pine weevil is an insect that kills the topmost shoot, and often causes the tree to have multiple stems and a round profile. In the forest, a white pine tree has a narrow head; and

the trunk is commonly free of live branches for a considerable portion of its length. Old forest trees have a broad and somewhat irregular head. The branches are horizontal and in regular whorls, usually of 5 each. Very old trees often become very irregular and picturesque. The trunk tapers gradually, and the tree often attains a height of 100 feet. Commonly it is from 70–80 feet tall, and has a diameter of 1–3 feet.

The **bark** of young trees is smooth and thin, green with a reddish-brown tinge overall, or brown in spots. On old trees, it is from 1–2 inches thick, very dark, and divided into broad, flat ridges by shallow fissures.

Leaves are in clusters of 5, flexible, 3–5 inches long, bluish-green but whitish on one side. The papery sheath at the base of the new needle clusters falls in late August.

The **cones** are 4–8 inches long, cylindrical and borne on a long stalk. They take 2 years to mature, and open to discharge the seed shortly after ripening

Young bark (left) and old bark (right).

in late August through September of the second season.

The **wood** is light in color and durable, except when in prolonged contact with moisture. It is soft, not heavy and is easily worked. The wood is used extensively for interior trim, doors, windows, cabinetmaking, sash and door manufacture, patternmaking, furniture, small building construction, interior and exterior finish, and boat planking.

Pine furniture is always popular in North America. Lumber from Maine is sold from Newfoundland to Washington state and south into Mexico. Lower grade boards have clear sections cut to size for sale. These clear short pieces may also be finger-jointed to create longer lengths of clear wood. Any part of a pine not making log grade is used for pulp. Ceiling tiles and paper are made from this pulp.

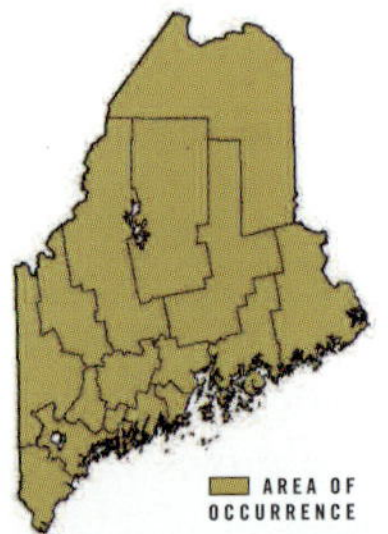

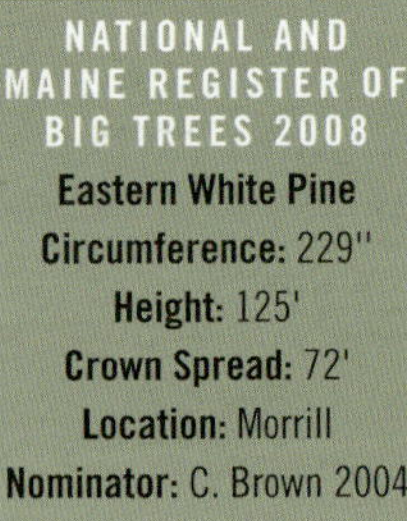

Eastern white pine leaves (needles) are 3–5 inches long and in clusters of 5.

White pine has been an important timber tree in Maine for more than 300 years.

R E D   P I N E *Pinus Resinosa* Soland.

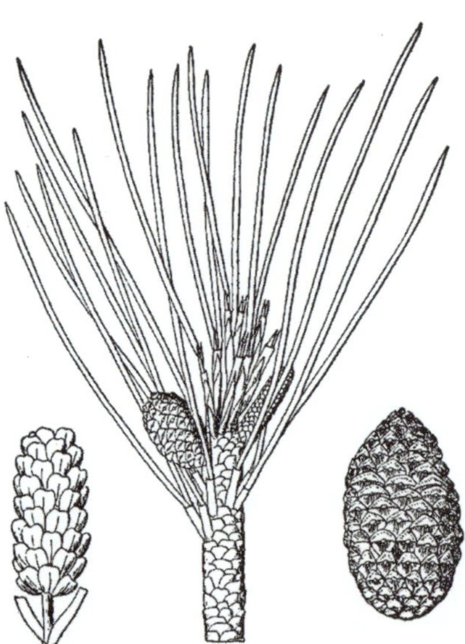

The red pine is named for its reddish-brown bark and pale red heartwood.

Red or Norway pine, though common, is found only locally throughout the state, growing on dry, rocky ridges, or light, sandy soil. Stands are usually scattered through forests of other species. The beautiful "Cathedral Pines" occur near Eustis.

Young trees often have branches extending to the ground and form a conical outline. Later, the head is rounded and picturesque. Branches are generally horizontal. It attains a height of 60–80 feet, and a diameter of 2–3 feet. The trunk is straight and tapers slowly. Red pine is not tolerant of shade.

The reddish-brown **bark** is divided into broad, flat ridges by shallow fissures.

The **leaves** are arranged in clusters of two. They are 4–6 inches long, dark green, soft and flexible. When doubled between the fingers, they break cleanly, at a sharp angle.

The **cones** are egg-shaped and are about 2 inches long. They lack prickles and are borne on short stalks. The base

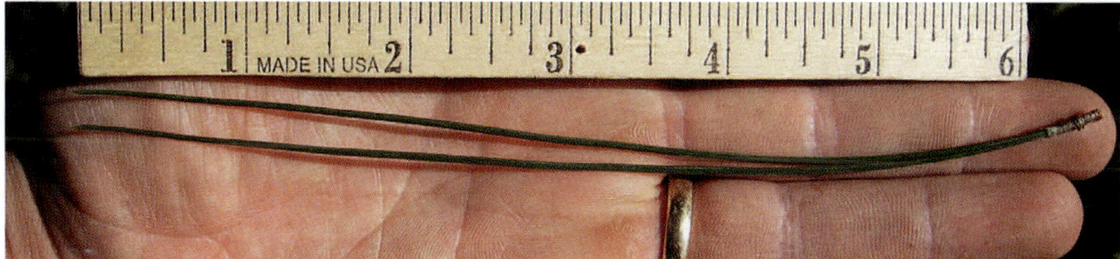

of fallen cones is hollow. They mature in the fall of the second season and usually remain on the branches until the following summer. Cones may be collected for seeds from September throughout the fall and winter, due to their gradual release of seed.

The **wood** is a little heavier and harder than white pine, close-grained, and fairly strong. It is used for lumber, poles, piles, building construction and pulp. It is treated readily with wood preservatives, and therefore is a locally-produced alternative to southern yellow pine. Older stands produce large, high-value poles.

Owing to the reddish bark and the pale red heartwood, the name "red pine" is appropriate. The name "Norway pine" refers to its original finding near Norway, Maine. Since it implies that the tree is foreign in origin, use of this name is discouraged.

The reddish-brown bark of the red pine is divided into broad, flat ridges by shallow fissures.

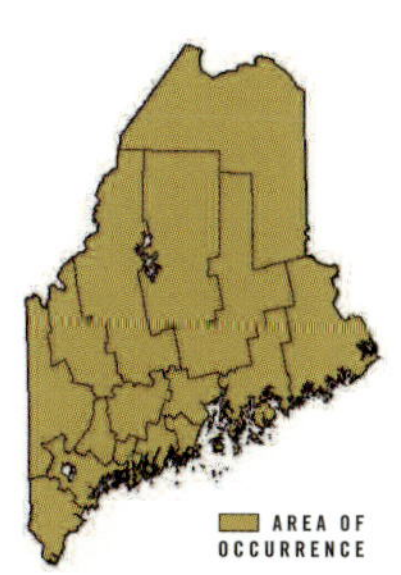

JACK PINE *Pinus banksiana* Lamb.

The cones of the jack pine usually remain closed for several years unless exposed to extreme heat, and often do not fall for 12–15 years.

Jack or gray pine grows on sandy, rocky, shallow acidic soils. It is known to occur naturally at Alamoosook Lake in Orland, Schoodic Point in Winter Harbor, Great Wass Island in Beals, Matagamon Lake, Cliff Lake, Lobster Lake, and in the areas south and west of Jackman.

The spreading **branches** are long and flexible, and form an open head symmetrical in outline. At maturity the tree is about 50–60 feet tall and 8–10 inches in diameter. Trees in the coastal populations tend to be much shorter and usually have a picturesque, gnarled look. **Cones** are often produced when the trees are only a few years old.

Jack pine cones are curved and persist on the tree for many years.

The **bark** is thin with irregular rounded ridges. It is dark brown with a slight tinge of red. The **leaves** are in clusters of two, and are ¾ to 1½ inches long. They are stout, yellow-green at first, dark green later, rather flat, and twisted at the base. The cones require 2 years to mature, are rather slender, 1½–2 inches long, lack a stalk and are curved. The scales have minute prick-les that are often deciduous. The cones usually remain closed for several years unless exposed to extreme heat, and often do not fall for 12–15 years.

The **wood** is moderately hard, heavy, and close-grained. It is used mostly for pulp; historically it was used for firewood and box boards.

Jack pines growing on the coast in eastern Maine are often stunted and gnarled.

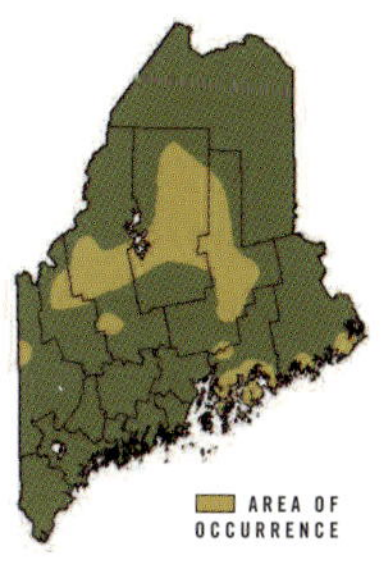

PITCH PINE *Pinus rigida* P. Mill.

Pitch pine wood is used for construction lumber, pulp and fire-starting "fat wood."

Pitch pine grows on sandy barrens or plains, and on gravelly soil of the uplands. It is quite common in the southern part of the state, on the sand plains near Brunswick and Oxford and on Mt. Desert Island.

Branches are horizontal, rigid, contorted and form an open crown. Pitch pine attains a diameter of 1–2 feet, and a height of only 30–40 feet. The trunk tapers rapidly and generally is straight. Often the tree produces cones when small. It is the only native pine that will resprout when damaged by such factors as fire.

The **bark** is rough, even on young stems and branches. On old trees, it is irregularly divided into continuous broad flat ridges, and is deep gray or reddish-brown.

The **leaves** are in clusters of three, and are 3–5 inches long. They are dark yellow-green and stiff, standing at right angles to the branch.

The **cones** require 2 years to mature, are 1½–3½ inches long, borne

Pitch pine cones have a sharp prickle at the end of each scale.

Pitch pine often has needles growing directly out of the trunk. This plus its clusters of 3 needles make it easy to recognize.

on short, hardly-noticeable stalks, and are often produced in clusters. A sharp, rigid, curved prickle is produced on the tip of each scale. The cones open gradually during midwinter. Seeds are released over a period of several years. Cones often remain on the trees 10–12 years. Fresh cones are used in wreath decorations.

The **wood** is moderately heavy, strong, hard and stiff. It is used for construction lumber, pulp and fire-starting "fat wood." In the past, considerable quantities of pitch and turpentine were obtained from this tree; these commodities were referred to as "naval stores," a term originally applied to the resin-based components used in building and maintaining wooden sailing ships. Today naval stores are used in the manufacture of soap, paint, varnish, shoe polish, lubricants, linoleum and roofing material.

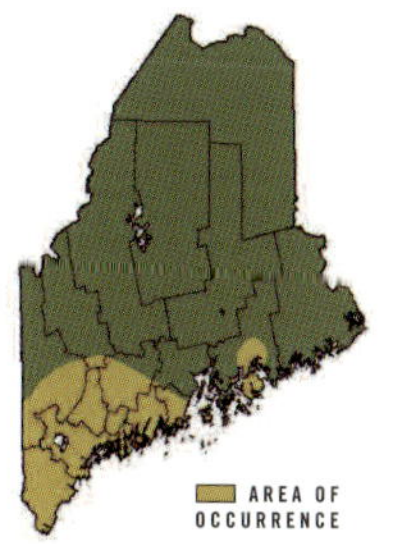

Scots (Scotch) Pine *Pinus sylvestris* L.

Scots pine is the most widely distributed pine in the world but is not native to Maine. A native of northern Europe and Asia, it grows naturally from Scotland almost to the Pacific Ocean and from above the Arctic Circle in Scandinavia to the Mediterranean. In parts of its native range, Scots pine grows to be a tall timber tree in dense stands. There are many strains of this species; the trees that have been planted in Maine often have very poor growth habits. This, plus its susceptibility to snow, porcupine and bird injury, makes it undesirable for timber production here. Scots pine will grow on very poor soils. Some strains are planted for Christmas trees, and it has been widely used in ornamental plantings.

The **bark** in the crown region of medium to large trees has conspicuous orange coloration. The lower bark of mature trees is gray to red-brown and has irregular ridges and furrows.

The **leaves** are needles in clusters of two. They are 1½–3 inches long, stout, stiff, twisted, dull blue-green with distinctive lines of stomata.

The **cones** are 1½–2 inches long and numerous, even on comparatively young trees; the scales are armed with small, blunt spines.

The **wood** is similar in character to red pine; however, due to its poor form, in Maine it is seldom used for lumber. It is occasionally used for pulp and fuel.

Scots pine, as the name suggests, is not native to Maine.

MAINE REGISTER OF BIG TREES 2008
Scots Pine   Circumference: 124"   **Height:** 60'   **Crown Spread:** 45'   **Location:** Falmouth

SPRUCE *The Important Distinctions*

	Black Spruce *Picea mariana*	**Red Spruce** *Picea rubens*	**White Spruce** *Picea glauca*	**Norway Spruce** *Picea abies*
NEEDLES				
COLOR	Blue-green	Dark yellow-green	Blue-green to dark green	Dark green
LENGTH	$\frac{1}{4}$–$\frac{1}{2}$ inches	$\frac{1}{2}$–$\frac{5}{8}$ inches	$\frac{1}{2}$–$\frac{3}{4}$ inches	$\frac{1}{2}$–1 inch
DESCRIPTION	Dull with waxy bloom	Very shiny	Dull with waxy bloom, strong, unpleasant odor when crushed	Shiny, sharp pointed
CONES				
LENGTH	$\frac{1}{2}$–1$\frac{1}{2}$ inches	1$\frac{1}{4}$–2 inches	2 inches	4–7 inches
RETENTION	Remain on tree for many years	Fall first year	Fall first year	Falls first year
SHAPE	Spherical	Wide in middle	Cylindrical	Cylindrical
SCALES	Stiff and rigid when ripe; margin irregularly notched	Stiff, with margin entirely or slightly notched	Flexible at maturity, margin entire	Stiff, irregularly notched
TWIGS				
COLOR	Yellow-brown to brown	Reddish to orange-brown	Light gray to yellow-brown	Orangish-brown
HAIRS	Short, rusty to black hairs; some hairs tipped with globose glands	Short, rusty to black hairs; tips lack glands	Without hairs	Without hairs (twigs droop from main branch)

Seed of all spruce is winged; cones are pendant; bare twigs are roughened by persistent leaf bases

BLACK SPRUCE *Picea mariana* (P. Mill.) B. S. P.

In the past, spruce beer was made by boiling the branches of the black spruce.

Black spruce occurs statewide; it grows on cool upland soils, but is more commonly found along streams, on the borders of swamps and in sphagnum bogs. It is also often found on the sandy soils of eastern Maine. It can grow to a height of 50–70 feet and a diameter of 6–12 inches, but is normally smaller than the maximum size. On a good site, it will grow rapidly. In sphagnum bogs, trees 50–80 years old may be only 6–8 feet tall and about one inch in diameter. The branches are short, pendulous and have a tendency to curve up at the ends. It forms an open, irregular crown. The lower branches often touch the ground, and root to form new trees. This method of reproduction is known as "layering."

The **bark** on the trunk is grayish-brown and the surface is broken into thin scales. The **leaves** are ¼–½ inches long, dull blue-green, blunt-pointed, flexible and soft to the touch.

The **cones,** which usually stay on the trees for many years, are ½–1½ inches long, ovoid, and become nearly spherical when open. The cone scales are stiff and have toothed margins.

The **twigs** have many hairs, some of which are tipped with glands. The inner bark is olive-green.

The **wood** is soft and light, but strong. It is used for pulp, framing and construction lumber, and planking. Historically, spruce beer was made by boiling the branches.

Black spruce cones persist on the tree for many years. Look for clumps of old, gray, weathered cones high in the tree.

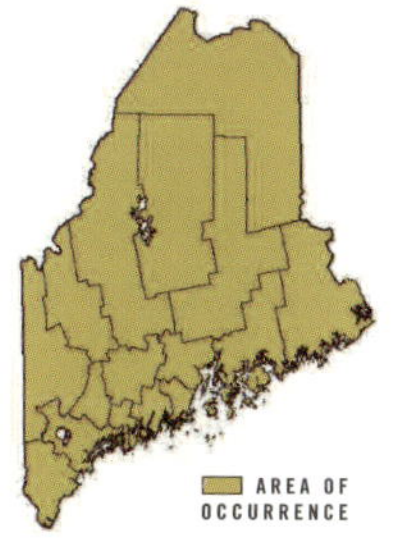

MAINE REGISTER OF BIG TREES 2008
Black Spruce Circumference: 47" **Height:** 66' **Crown Spread:** 20' **Location:** Camden

RED SPRUCE *Picea rubens* Sarg.

Red spruce is one of our most valuable trees for the production of building lumber.

Red spruce is commonly found throughout the state. It grows on well-drained, rocky upland soils, and particularly on the north side of mountain slopes where it may be the major species present. The spreading branches form a somewhat conical, narrow head in young trees. The trunk is long, with a slight taper. It grows to considerable size, and is capable of attaining a height of 60–80 feet and a diameter of 1–2 feet, but occasionally exceeds these measurements. Red spruce is shade-tolerant and will become established in the understory of mixed stands.

The **bark** on mature trees is thick and is broken into thin, reddish-brown scales of irregular shape. The **leaves** are dark green, often with a yellow tinge, and are very shiny. They are about ½ inch long, sharp-pointed, stiff, prickly to the touch, and point toward the tip of the branch. The **cones** are oblong and usually 1½–2 inches long. When ripe, they are reddish-brown and quite shiny. The cone scales are stiff like the

black spruce, but the margins are generally without conspicuous notches. The cones begin to drop in autumn or early winter, and are all gone from the branches by the next summer.

The **twigs** have hairs, none of which have a gland at the tip. The inner bark is reddish-brown. The **wood** is fairly soft, light, close-grained and strong, but is not as durable as pine when exposed to the weather.

Red spruce is one of our most valuable trees for the production of building lumber. It is used for joists, sills, rafters, pilings, weir poles and heavy construction timbers. It is a principal wood used in the manufacture of paper pulp, and is valuable for the sounding boards of musical instruments. Pitch for spruce gum is obtained largely from this tree.

Red spruce is the characteristic tree of the "Acadian forest" of northern New England and the Canadian Maritimes.

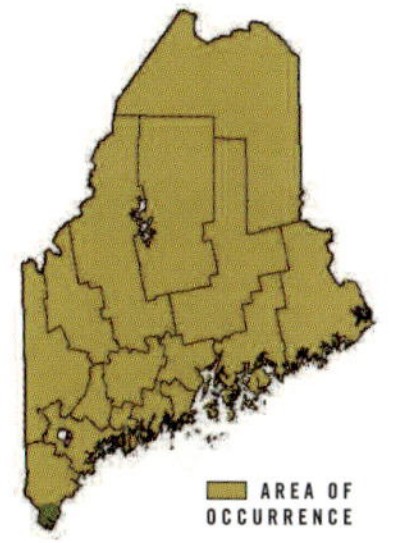

MAINE REGISTER OF BIG TREES 2008
Red Spruce Circumference: 103" Height: 87' Crown Spread: 35' Location: T15 R9 WELS

WHITE SPRUCE *Picea glauca* (Moench) Voss

The wood of the white spruce is used for pulp, paddles, oars, piano sounding boards and dimension lumber, while its cones are used to make decorative wreaths.

White or cat spruce occurs statewide except in York county. It is widely distributed, but not as abundant as red spruce. It grows on shallow, rocky sites from the coast to the tree line in the mountains, and is also commonly found in old pastures and on cleared land. It does not tolerate shade and does not grow as an understory tree. The long and rather thick branches, densely clothed with stout, rigid lateral branches, are curved upward and form a somewhat open, irregular head with a broad base. It commonly grows to a height of 60–90 feet and to a diameter of 2 feet.

The **bark** on old trees has light gray, plate-like scales, which are thin and irregular, with a somewhat brownish surface. Younger trees have smooth, light gray bark.

The **leaves** point straight out from the branch. On the lower half of the twig the leaves are often bent upward in such a manner as to bring them all on the upper side. They are pale blue-

green at first, later becoming a dark blue-green. The foliage emits a peculiar and characteristic odor, which is a ready means of distinguishing it from the other spruce species and is the reason for the alternate name.

The **cones** are slender, cylindrical, pale brown and shiny when ripe, and usually about 2 inches long. They ripen in August and September, and may be collected for seed until October. Cones usually fall off the first year. The cone scales are thin and flexible, so that they give easily when the cone is clasped in the hand. The **twigs** are without hairs. The inner bark is silvery and glistens.

The **wood** is fairly light, soft, finishes well and is moderately strong. It is used for pulp, paddles, oars, piano sounding boards and dimension lumber, while its cones are used to make decorative wreaths. It shouldn't, however, be used as a Christmas tree; when it is brought indoors, the reason for its nicknames—cat spruce and skunk

White spruce cones are cylindrical and the scales can be easily broken apart. This distinguishes it from red and black spruce, which have globe or egg-shaped cones with stiff scales.

spruce—become evident. White and black spruce produce long, tough, pliable roots which were used by American Indians to tie together pieces of birch bark for canoes and other purposes.

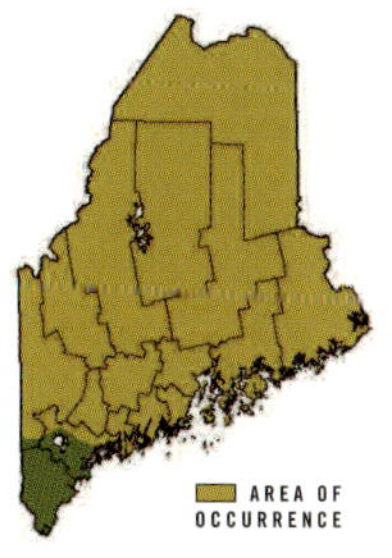

Norway Spruce *Picea abies* (L.) Karst.

A native of Europe, Norway spruce is of great economic importance in its natural range. In Maine it is commonly planted both in forest plantations and as an ornamental tree. It rarely reproduces in the wild.

It is very symmetrical and graceful in its growth habit; open-grown trees often carry branches clear to the ground. The tips of branches on larger trees have an upward sweep; and lateral branchlets are long and pendent. Norway spruce grows more rapidly than any of our native species of spruce, and has been frequently planted for pulpwood, particularly in old fields in Aroostook County. It is very susceptible to attack by the white pine weevil.

The **bark** of younger trees is reddish-brown; older trees have grayish bark with flaking scales. The **cones** are large, 4–7 inches long, and cylindrical with stiff, notched scales. The **leaves** are ½–1 inch long, deep shiny green, four-sided in cross section and slightly flattened. The needles lack the tendency to bend upward on the twigs as in white spruce. **Twigs** are orangish-brown and without hairs.

In its native Europe, Norway spruce is a very important lumber and pulpwood species. In Maine, the **wood** is primarily used for pulp and occasionally for lumber.

As its name implies, Norway spruce is not native to Maine.

BLUE SPRUCE *Picea pungens* Engelm.

A native of the Rocky Mountain region, blue spruce will grow on a variety of sites and tolerate a wide range of growing conditions. These factors, plus the striking color of its foliage, contribute to its popularity as an ornamental species, particularly in the East where it is planted as a decorative tree. It does not readily become naturalized in Maine, and therefore is not likely to be found growing in forest settings. It can grow to be a large tree 1–2 feet in diameter and to about 80 feet in height. It is pyramidal in shape.

Foliage coloration varies from silvery-blue to blue-green; the intensity of blue varies between individual specimens. The **bark** is gray to red-brown and scaly. The **leaves** are ½–1½ inches long, stiff, very sharp-pointed, and strongly incurved and covered with a waxy coating that gives the blue color.

Blue spruce is not native to Maine and is not likely to be found growing in forest settings.

Cones are light brown, oblong, 2½–4 inches long, with thin, flexible, notched scales. The twigs are stouter than the other spruces, hairless and tan.

Wild trees growing in the Rocky Mountains seldom have the intense coloration of the cultivated varieties planted here. Even in its native range, the wood is not often used commercially because of its limited availability and its tendency to be brittle and full of knots.

BALSAM FIR *Abies balsamea* (L.) P. Mill.

Balsam fir is the most abundant tree in the state.

Balsam fir occurs statewide and is the most abundant tree in the state. It is frequently found in damp woods and on well-drained hillsides, and often occurs in thickets. The tree normally forms a sharp spire to a height of 60–70 feet and grows to 12–20 inches in diameter. On young trees, the branches are horizontal, slender, and produced in regular whorls to form a strikingly symmetrical crown. In old age, the top is often slim, regular and spire-like.

The **bark** on young trees is pale gray, smooth, thin and has prominent blisters that are filled with a resinous liquid known as "Canada balsam." On old trees the bark gets rougher and blisters are absent.

The aromatic **leaves** are about 1 inch long, dark green, and shiny above with 2 rows of white stomata below. The tips are occasionally notched. On branches in full sun, leaves turn up, but on lower branches they spread out at right angles to the branch, giving it a flattened appearance.

Like all true firs, balsam fir cones point upward and disintegrate when they are mature.

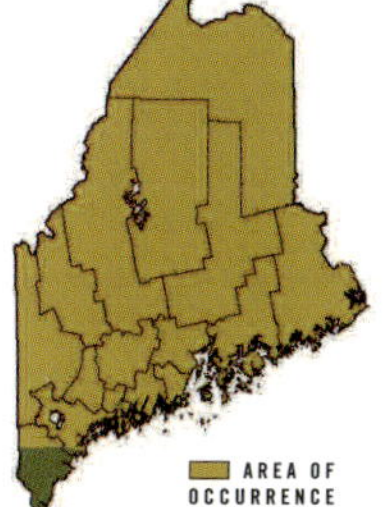

The **cones** are 2–4 inches long, erect and dark purple before maturity. Cones ripen in August and September of the first year, and disintegrate shortly thereafter, leaving only the central spike-like stalks. The twigs are smooth after the leaves have shed. Winter buds are covered with clear resin.

The **wood** is soft, light and moderately limber. It is sawed into dimension lumber chiefly for light and medium building construction, and is used extensively for pulp. Balsam fir is favored for Christmas trees and greens. Each fall many tons of branch tips are collected for making Christmas wreaths. In the past, the branches were steamed in a retort to produce oil of balsam. Also, the clear pitch formed in the blisters of relatively young bark was used to mount microscope slides and to attach theatrical costumes to bare skin.

The smooth bark with resin blisters distinguishes balsam fir from the rest of our conifers.

EASTERN HEMLOCK *Tsuga canadensis* (L.) Carr.

The wood of the Eastern hemlock is used for framing, sheathing, roof boards, timbers, bark mulch and pulp.

Eastern hemlock is found in scattered stands in nearly every part of the state. Best growth is attained on moist, cool sites. It generally attains a height of 60–70 feet, and a diameter of 2–3 feet. The terminal shoot droops and bends away from the prevailing winds, quite often toward the east. The trunk usually tapers rapidly from the base. This species can withstand considerable shading.

The **bark** is divided into narrow, rounded ridges covered with thick scales, and varies in color from cinnamon-red to gray. Inner bark exposed by cuts or bruises shows a purplish tinge.

The **leaves** are flat, tapering, generally rounded at the apex, from ⅓–⅔ inch long, with a distinct short petiole and so arranged that the twig appears flat. Leaves become progressively shorter towards the tip of the twig. They are dark yellow-green with a lustrous upper surface, and a whitish undersurface.

The **cones** are about ¾ inch long, oblong, light brown, pendant and suspended on short, slender stalks. Cones mature during the first autumn and generally remain on the branches until the next spring. Seeds are winged and fall during the winter. The **twigs** are very fine, limber and are not pitchy.

The **wood** is coarse, brittle when very dry, light, strong and difficult to work as it is likely to separate at one or more of the annual growth rings. It is used for framing, sheathing, roof boards, timbers and pulp. The bark was once valuable for tanning but has been replaced by chemicals; now it is prized for its purple color when made into mulch.

When cut with a knife, Eastern hemlock bark will show a purple color.

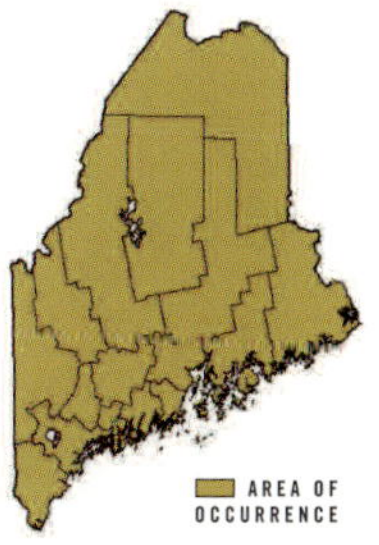

MAINE REGISTER OF BIG TREES 2008
Eastern Hemlock Circumference: 127'' **Height:** 88' **Crown Spread:** 32' **Location:** North Yarmouth

TAMARACK *Larix laricina (Du Roi)* K. Koch

Tamarack, eastern larch or hackmatack is most commonly found in cool, swampy places, although it also grows on well-drained soil. It is found in scattered stands throughout the state. It can grow rapidly and is not tolerant of shade.

In the forest, the tree grows to a height of 50–60 feet and a diameter of 20 inches. It has a regular, narrow, pyramidal head with small, stiff horizontal branches.

In northern Maine, the name "juniper" is quite commonly applied to this tree, but since juniper is the true name of another tree, its use for tamarack is discouraged.

Tamarack is our only native conifer that sheds all its leaves every fall.

TAMARACK

> **NATIONAL AND MAINE REGISTER OF BIG TREES 2008**
> **Tamarack**
> **Circumference:** 143"
> **Height:** 92'
> **Crown Spread:** 31'
> **Location:** T13 R8 WELS

The **bark** separates on the surface into small, thin, irregular reddish-brown scales.

The **leaves** are linear, about 1 inch long, triangular in cross section, and borne in clusters of 8 or more on spurs, except on elongating new shoots, where they occur singly. They are bright green and turn a beautiful yellow just before they fall. Tamarack provides some of the last color of the fall, as its needles turn color after most trees have already shed their leaves. It is our only native conifer that sheds all its leaves every fall.

The **cones** are small, nearly spherical, about ¾ inch long, light brown, and borne erect on stout stems. They open in fall to liberate the small winged seeds and usually remain on the tree until the following year.

The **wood** is rather coarse-grained, hard, heavy and strong, with durable heartwood. It is used for planking, timbers, ties, poles, signposts, pilings and pulp. Historically, tamarack knees (the buttresses formed by large roots) were used in shipbuilding. Tamarack was also used for mud sills in home construction.

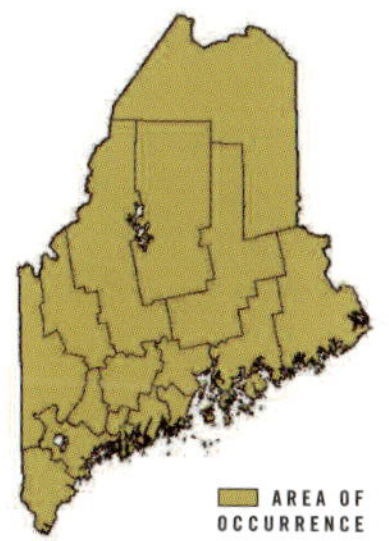

On older twigs, tamarack needles occur in clusters of up to 50 on short spur branches.

ATLANTIC WHITE CEDAR
Chamaecyparis thyoides (L.) B. S. P.

While Atlantic white cedar wood is of limited use, small trees are cut for fenceposts and shavings are used for pet bedding.

Atlantic or coast white cedar is found in bogs or low areas along ponds or streams. It has a scattered distribution from the mid-coast south. In Maine it rarely reaches a height of over 40 feet. The short branches come out from a gradually tapering trunk, giving the tree a conical appearance. The twigs are only slightly flattened.

The **bark** is fibrous, grayish to reddish-brown, often with twisted spirals; on young trees it is easily pulled off in strips.

The **leaves** are bluish-green, scale-like, and arranged in somewhat fan-shaped clusters. When crushed, they give off an aroma.

The **cones** are small, round, smooth and purplish before maturity, about ¼ inch in diameter with tack-like scales. They persist through the winter, but are inconspicuous.

The **wood** is light, close-grained, strongly fragrant, and light brown tinged with red. It is brittle and therefore of limited use, though small trees are cut for fenceposts. The shavings are used for pet bedding.

Atlantic white cedar is rare in Maine and occurs only in a few isolated bogs in the south and mid-coast.

NORTHERN WHITE CEDAR
Thuja occidentalis L.

Cedar has emerged as a viable alternative to pressure-treated wood.

Northern white cedar or eastern arborvitae is generally found in swamps, along streams, on mountain slopes and in old pastures where the soil is moist. Dense stands are widely distributed statewide. It is most abundant in the northern and eastern sections, and grows best on alkaline soils. It is widely used as an ornamental. The head is compact, narrow and pyramidal. The branches are horizontal, short and turned upward. Trees grow to 60 feet in height and to 3 feet in diameter. The trunk is often strongly buttressed.

The **bark** has shallow fissures, which divide it into flat narrow ridges. It is reddish-brown and often tinged with orange.

The **leaves** are opposite or two-ranked, usually only about 1/8 inch long, scale-like, blunt, and so arranged as to make the small branches flat in

Northern white cedar cones are about $^1/_2$ inch long and often occur in large numbers.

shape. They have a pleasant aroma and a rather pleasing taste, and are a major source of food for deer in the winter.

The **cones** are erect, small, about ½ inch long, with only a few pairs of scales. They mature in one season. The seed is small and winged.

The **wood** is soft and light, coarse-grained, brittle, has very durable heart-wood and a fragrant odor. It is used primarily for shingles, slack cooperage (barrels for dry, semi-dry or solid products), poles, posts and rustic fencing; and it is sawed into lumber for hope chests (since the wood is said to repel moths), siding, canoes and boats. More recently, cedar has emerged as a viable alternative to pressure-treated wood. Naturally weather-resistant, it is used for decks, post and rail fencing, outdoor furniture, roof shakes, and pelt stretchers.

EASTERN REDCEDAR *Juniperus virginiana* L.

In Maine, Eastern redcedar is not sufficiently plentiful to be of commercial importance.

Eastern redcedar is not common in Maine. It grows on poor soils, gravelly slopes, rocky ridges and on moist, sandy ground. It is found intermittently in southern Maine and in Bridgton, Porter, Denmark and West Gardiner. It gets the name "redcedar" from the color of the heartwood.

It is variable in its habit. Young trees have slender horizontal branches and a narrow, compact, conical head. The crown of old trees becomes broad and rounded. In Maine, trees attain a diameter of 8–12 inches, and a height of 30 feet.

Eastern redcedar invades old pastureland and quickly dies out when other trees begin to shade it.

The **bark** on the trunk is light brown, tinged with red; it separates into long, narrow shreds on old trees.

The **leaves** are scale-like, overlapping, about 1/16 inch long, dark green, and remain on the tree 5–6 years, growing hard and woody the third season. Branchlets appear square in cross section. Current growth and vigorous shoots contain sharp-pointed, awl-shaped leaves—the so-called "juvenile" growth.

The **fruit** is berry-like, globose, with 1–2 seeds, pale green at first, dark blue when ripe, and is about the size of a small pea.

The **wood** is brittle, fine-grained, light, easily worked, durable, and very aromatic. The heartwood is a dull red. It is valuable for fence posts and paneling for moth-proof closets, but in Maine it is not sufficiently plentiful to be of commercial importance. The shavings are used as bedding for pets.

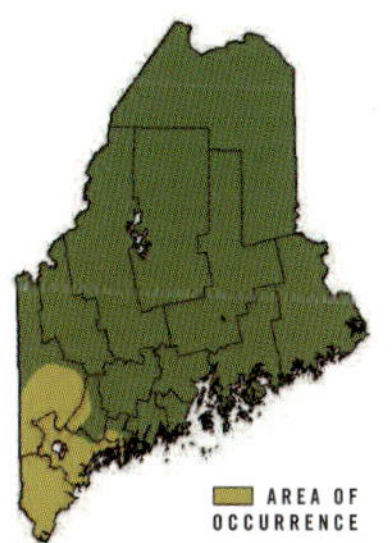

COMMON JUNIPER *Juniperus communis* L.

Common juniper is found primarily as a shrub in pastures and open spaces on shallow, rocky soil. It occurs infrequently, primarily in the southern half of the state. It is occasionally found as a tree. Specimens up to 25 feet in height have been recorded, but are extremely rare.

The **bark** is grayish-brown and occurs in thin, longitudinal, shredded layers. The inner portion has a reddish tinge. The **leaves** occur in whorls of three. They are sharp, stiff, dagger-like and persist for several seasons. They are ¼–¾ inch in length. The upper surface is concave and marked with a broad, white line. The underside, which due to the bending of the twigs usually appears uppermost, is dark green.

The **fruit** is dark blue, covered with a thin bloom and is slightly smaller than a pea. Fruits remain on the trees during the winter, and have a strong resinous taste. The fruit is usually found only on select trees since male and female flowers are generally produced on separate trees. This trait is common to most junipers.

The **wood** is hard, close-grained and very durable. The heartwood is light brown. Large stems make long-lasting fence posts if the bark is removed.

Juniper shavings can be used for pet bedding. In Europe, the fruits are used to make gin.

Common juniper is usually found as a shrub rather than as a tree.

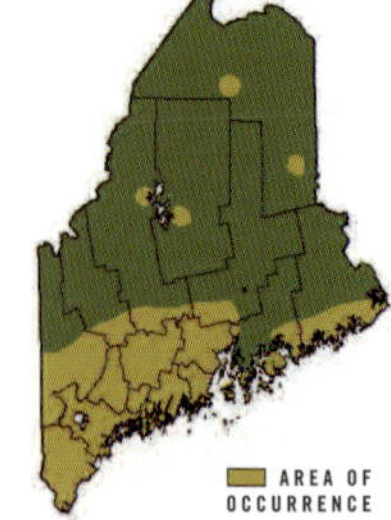

Horses were still commonly used to haul logs in the Maine woods until the 1950s.

BROADLEAVES

Often referred to as "hardwoods", broadleaf trees belong to the group of plants known as the angiosperms or flowering plants. Almost all of Maine's hardwood species are deciduous, meaning they loose all of their leaves each autumn and grow new ones in the spring. Maine has over 50 species of native hardwood trees; only about half of these are considered important timber trees. The name "hardwood" is somewhat misleading since some hardwood species have wood that is soft.

Photo location: T1 R6 WELS

POPLARS & ASPENS *The Important Distinctions*

	Quaking Aspen *Populus tremuloides*	**Bigtooth Aspen** *Populus grandidentata*	**Balsam Popular** *Populus balsamifera*
BARK			
TEXTURE	Smooth in younger trees, often with horizontal bands of circular wart-like outgrowths	Smooth in young trees; furrowed in older trees	Smooth or roughed by dark outgrowths; older trees furrowed with scaly ridges
COLOR	Light or grayish-green	Dark or olive green	Reddish-brown on younger trees
TASTE	Very bitter	Not bitter	Not bitter
LEAVES			
LENGTH	1½–3 inches	3–4 inches	3–5 inches
SHAPE	Circular	Broad egg-shaped	Egg-shaped
MARGIN	Finely toothed	Coarsely toothed	Finely toothed
SURFACE	Shiny upper, not rusty beneath	Not shiny upper, not rusty beneath	Very shiny upper, rusty beneath
PETIOLE	Flattened	Flattened	Flattened
BUDS			
TEXTURE	Not sticky; shiny	Not sticky; dull	Very sticky; shiny
SHAPE	Conical	Broad egg-shaped	Egg-shaped
SCALES	No hairs	Covered with white hairs	No hairs
ODOR	Not fragrant	Not fragrant	Sweet balsam fragrance

The pith of poplar twigs is star-shaped in cross section. Poplars belong to the willow family and resemble willows in flower and fruit characteristics. The nodding, "woolly bear" caterpillar-like staminate and pistillate catkins are borne on different trees. They open before the leaves are out and are conspicuous in the early spring. Poplars, like willows, have a transcontinental range. They can be propagated very easily from cuttings.

QUAKING ASPEN *Populus tremuloides* Michx.

Quaking aspen, popple or trembling aspen is found statewide and is an abundant, rapid-growing tree occurring in either pure stands or in mixture with other species. It is found on many different kinds of soil, but makes the best growth on sandy, moist soils. Frequently it is the first species, with paper birch, to become established following heavy cuttings or burns. Intolerant of shade, it does not persist in dense woods. It is a graceful tree with slender branches that are far apart and often contorted. It has a round and narrow head. It grows to a height of 60–75 feet and a diameter of 10–16 inches.

The **bark** is smooth, often roughened by horizontal lines of wart-like outgrowths. It is a pale green with dark brown patches. The pale green areas feel waxy when rubbed. Bark on old trees is ash gray and dark at the base where it is divided into broad, flat ridges. It has a very bitter taste similar to quinine.

The **leaves** are alternate, rounded and short-pointed, with finely rounded teeth; dark green and shiny above and 1½–3 inches long. The flattened petiole causes the leaves to tremble in a breeze, resulting in a rustling sound.

The **flowers** are in catkins that appear before the leaves. The **fruit**, which ripens about June, is a capsule. The seeds are very small, light and cottony, and are carried long distances by the wind. The **buds** are dark brown, have a varnished appearance and may be slightly sticky. Flower buds are usually larger than the leaf buds.

The **wood** is close-grained, soft and rots very easily. It is used increasingly for trim, lumber, pallets, and for the manufacture of oriented strand board, landscape ties, plywood, core stock and expendable turnery items. It is used extensively for pulp. In the past, it was ground up and cooked for cattle feed. Sometimes referred to as "biscuit wood," it was also used as firewood for cooking.

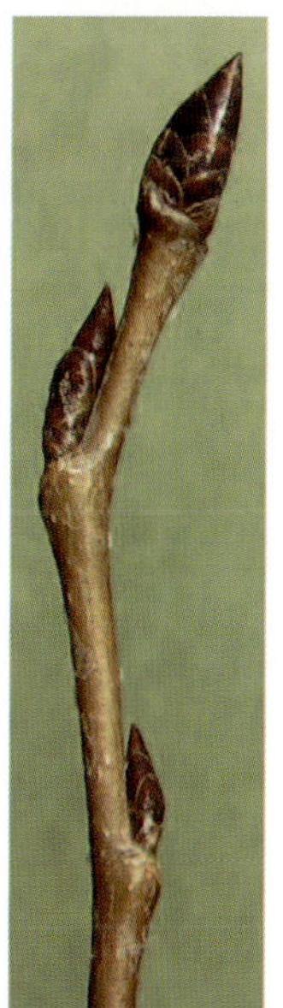

Above left: young bark.
Above right: old bark.
Left: Quaking aspen buds are dark brown and very shiny.

MAINE REGISTER OF BIG TREES 2008
Quaking Aspen
Circumference: 43"
Height: 69'
Crown Spread: 21'
Location: Richmond

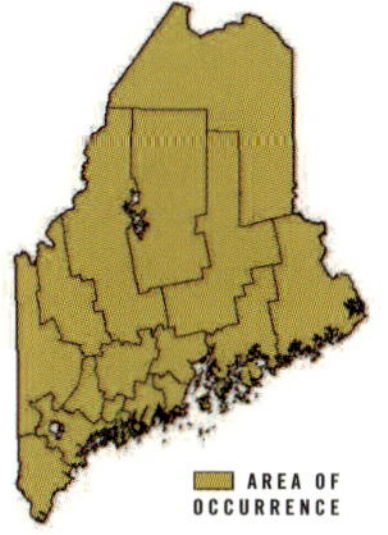

BIGTOOTH ASPEN *Populus grandidentata* Michx.

Bigtooth aspen, poplar or popple occurs statewide and commonly grows with quaking aspen. It is a rapid grower in various soils and in different situations. It grows best in a rich, sandy and fairly moist soil. It is more shade-tolerant, and therefore more competitive, than quaking aspen and grows with other species in either scattered or small groves. Bigtooth aspen tends to have better form than quaking aspen. It attains a height of 60–80 feet and a diameter of 10–20 inches.

When first emerging in spring, the bigtooth aspen leaf is a distinctive silvery-green.

The **bark** is smooth, and olive to gray-green. At the base of old trees, it is dark and divided into broad, irregular, flat ridges.

The **leaves** are alternate, 3–4 inches long, broadly egg-shaped in outline, and have a dark green upper surface. When first emerging in spring, they are a distinctive silvery-green. The edges are coarsely and irregularly toothed. The petiole, or leaf stalk, is flat.

The **flowers** are in catkins, and appear before the leaves.

The **fruit** ripens in May about the time the leaves begin to come out. The seeds are small, light and are carried long distances by the wind. The **buds** are dull gray, slightly hairy and not sticky.

The **wood** is like that of the quaking aspen and is used for the same purposes, as well as rails for apple-picking ladders.

NATIONAL AND MAINE REGISTER OF BIG TREES 2008
Bigtooth Aspen Circumference: 151"
Height: 76' Crown Spread: 45' Location: Appleton

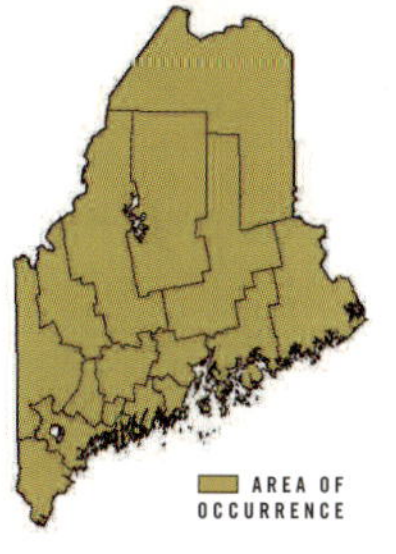

BALSAM POPLAR *Populus balsamifera* L.

Balsam poplar, or Balm-of-Gilead, inhabits the borders of swamps and the low bottomlands along rivers throughout the state, except in York County. It gets its name from the fragrance of the resinous, sticky buds.

The tree is somewhat different from the two preceding poplars. The branches are stout, erect, more or less contorted at the ends and form an open, rather narrow head. It reaches a height of 30–70 feet, and a diameter of 15–30 inches.

The **bark** on young trees is smooth, or sometimes roughened by dark outgrowths, and is greenish to reddish-brown. On the trunk of old trees, it is gray and separated into broad, rough ridges.

The balsam poplar gets its name from the fragrance of the resinous, sticky buds.

The **leaves** are alternate, ovate, 3–5 inches long and 2–3 inches wide. They are deep dark green and shiny on the upper surface, light green and usually with rusty blotches on the under side. The edges are lined closely with small, rounded teeth. The petioles are round in cross section. In late summer the entire tree can have a rusty appearance.

The **flowers** are in catkins that appear early in spring just before the leaves.

The **fruit** ripens the end of May or early in June. Each seed is attached to a cottony mass, so that it is often carried long distances by the wind.

The **wood** is somewhat like that of quaking and bigtooth aspen, but it is not as strong. The wood is prone to decay while growing. Larger logs are sawed into landscaping ties. OSB—oriented strand board, a structurally engineered wood product—can include a small percentage of balsam popular.

Balsam poplar has large sticky buds that have a sweet fragrance.

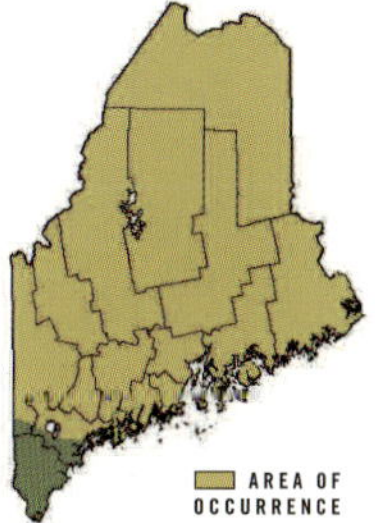

MAINE REGISTER OF BIG TREES 2008
Balsam Poplar Circumference: 99'' **Height:** 93' **Crown Spread:** 50' **Location:** Yarmouth

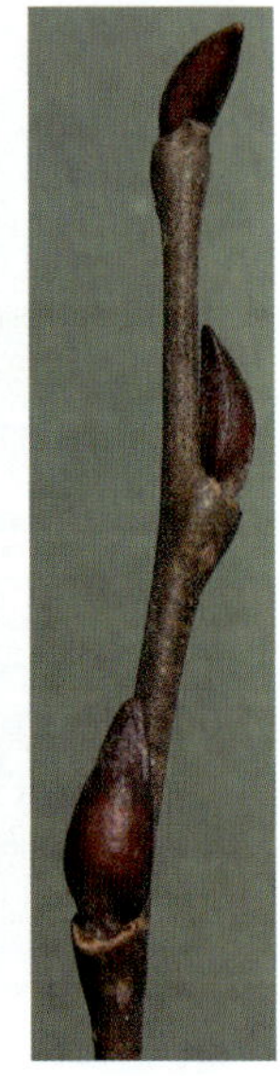

BLACK WILLOW *Salix nigra* Marsh.

Black willow occurs primarily in southern and western Maine. It grows to a height of 45–65 feet, and is found along streams and ponds. The stout, upright, spreading branches give the tree a broad, irregular outline. It is probably our largest native willow. The **bark** on old trees is shaggy and dark brown. The **leaves** are very narrow, sometimes sickle-shaped, finely-toothed, 3–6 inches long and green on both sides. The **wood** is soft, light, weak and is used occasionally for farm lumber and pulp.

Top right: A typical willow twig.

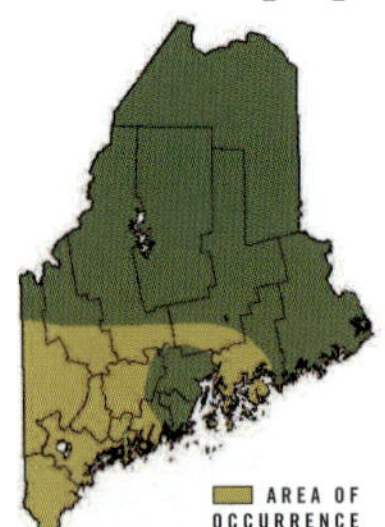

MAINE REGISTER OF BIG TREES 2008
Black Willow   Circumference: 266''   **Height:** 84'
Crown Spread: 64'   **Location:** Yarmouth

Much of Maine's accessible, large white pine had been cut by the 1850s. After that, spruce, seen here, became an increasingly important timber tree.

MAPLES* *The Important Distinctions*

	Red Maple *Acer rubrum*	**Sugar Maple** *Acer saccharum*	**Silver Maple** *Acer saccharinum*
BARK			
TEXTURE	**Older trees:** ridged and broken into plate like scales. **Young trees:** smooth.	**Older trees:** rough and deeply furrowed. **Young trees:** smooth and slightly fissured.	**Older trees:** somewhat furrowed, separates into thin plates. **Young trees:** smooth.
COLOR	Dark gray on older trunks, light gray on young trees	Gray on older trunks, light gray on young trees	Dark gray on older trunks, gray on young trees
LEAVES			
LOBES/SHAPE	3–5 lobes. sides of terminal lobe converge, notches between lobes V-shaped	3–5 lobes, sinuous, sides of terminal lobes flare outward, notches between lobes rounded	5 lobes, rarely 3; lobes long and narrow like fingers on a hand
MARGIN	Irregularly doubly toothed	Sparsely toothed	Irregularly and sharply toothed
SURFACE	Under-surface slightly white	Under-surface pale green	Under-surface silvery white
FLOWER			
APPEARANCE	Scarlet or yellow-red, appears before the leaves	Greenish-yellow, appears with the leaves	Greenish-yellow or pinkish, appears long before the leaves
BUDS			
LENGTH	Terminal bud—⅛ inch	Terminal bud—¼ inch	Terminal bud—⅛ inch
SHAPE	Blunt-pointed, as long as broad	Sharp-pointed, many scales showing	Blunt-pointed, slightly ridged
COLOR	Dark red	Purplish-brown to gray	Bright red above, green below
FRUIT			
SHAPE	Paired, slightly divergent	Paired and slightly divergent	Paired, but with one usually abortive
SEED BODY	Oval in outline	Round	Football-shaped
WING	Reddish; ¾ inch long	1 inch long	Strongly divergent, 2 inches long and hooked

	Striped Maple *Acer pensylvanicum*	**Mountain Maple** *Acer spicatum*	**Norway Maple** *Acer platanoides*
BARK			
TEXTURE	Marked with whitish stripes running lengthwise on trunk	Smooth when young; shallowly furrowed when older	Smooth when young; regularly furrowed on older trees
COLOR	Reddish-brown or dark green	Reddish-brown to gray	Gray when young, gray-brown on older trees
LEAVES			
LOBES/SHAPE	3 lobes, shaped; like a duck's foot, thin	Usually 3 lobes, sometimes 5	5–7 lobes, blade wider than tall, stems exude a milky sap when broken
MARGIN	Edges finely and sharply-toothed	Coarsely-toothed	Sparsely toothed
SURFACE	Under-surface pale green; pubescent	Prominently sunken veins on the upper surface	Very dark green; some cultivars red or deep purple
FLOWER			
APPEARANCE	Bright yellow, appears after leaves are full grown	Yellow-green in long clusters after the leaves are full grown	Yellow-green, appear before the leaves
BUDS			
LENGTH	Terminal bud—½ inch	Terminal bud—¼ inch	>¼ inch
SHAPE	Distinctly stalked with 2 scales showing	Slender and pointed, slightly stalked	Turban-shaped, blunt-pointed, large scales
COLOR	Bright red	Green to red	Green to purple
FRUIT			
SHAPE	Paired and moderately divergent	Paired, slightly divergent, ascending clusters	Paired, flattened
SEED BODY	Large smooth depression in seed body	Wrinkled depression on seed body	Flattened
WING	Reddish-brown; ¾ inch long	Slightly divergent; ½ inch long	Strongly divergent, leathery; 2 inches long

*Key does not include boxelder. Boxelder (page 84–85) is the only maple in Maine with compound leaves.

RED MAPLE *Acer rubrum* L.

Red maple—also known as soft, white or swamp maple—occurs throughout the state. A rapid grower and the most abundant of the maples, it is typically found in swamps and poorly drained sites, but also occurs elsewhere. The red maple is a medium-sized, slender tree that becomes 50–60 feet high, and 1–2 feet in diameter. The branches are upright, forming a somewhat narrow head. Usually the trunk is not divided.

The **bark** on young trees is smooth and light gray. On old trunks, it is dark gray, ridged and broken into plate-like scales.

Red maples produce bright red flowers followed by abundant seeds in the springtime.

The **leaves** are opposite, 3–5 inches long, with 3–5 lobes and margins that are irregularly double-toothed. The upper surface is light green; lower surface is white. The sides of the terminal lobe converge toward the tip; and the notches between lobes are V-shaped. In fall, the leaves turn scarlet and orange.

The **flowers** are produced in clusters on stalks before leaf buds open. Males are yellowish-red while females are bright scarlet. The red maple is one of the first trees to flower in spring.

The **fruit** is winged, ripens in spring or early summer, and germinates as soon as it falls. Wings are only slightly divergent, about ¾ inch long. The seed body lacks a depression.

The **twigs** are straight, stiff, do not have a rank odor when broken, and are red on both surfaces. Buds are red and often clustered.

The **wood** is close-grained, heavy, moderately strong, easily worked but not durable, although it will take a good polish. It is used mainly for pulp and firewood, but also for pallets, furniture stock, canoe paddles and turnery products. As sugar maple becomes more expensive, more mills are using red maple. It is also commonly used for landscape plantings.

MAINE REGISTER OF BIG TREES 2008
Red Maple Circumference: 183" **Height:** 69'
Crown Spread: 67' **Location:** Richmond

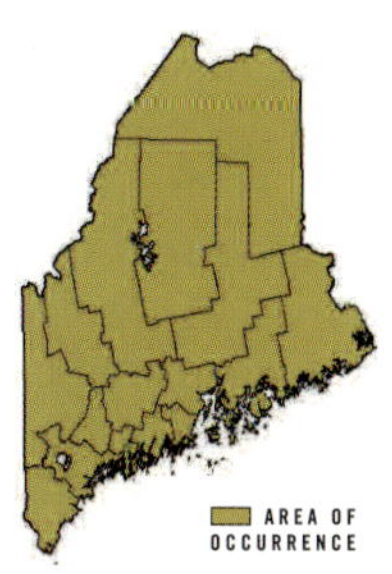

SUGAR MAPLE *Acer saccharum* Marsh.

Sugar, rock or hard maple is found abundantly throughout the state on moist, rocky slopes, but grows best on moist, upland soils. In the forest, it grows to 60–70 feet and a diameter of 20–30 inches. The top is short and spreading. In the open, the branches begin 8–10 feet up, forming an egg-shaped head when the tree is young and a broad, rounded top when older. It makes an attractive street or ornamental tree, but it is sensitive to road salt. Maple sugar and syrup are made largely from the sap of this tree, although sugar is present in the sap of all maples.

Historically, sugar maple was used to make parts for sleighs, sleds, pungs (low, one-horse box sleighs) and buggy shafts.

Bark on young trees and large branches is smooth or slightly fissured and pale. Some trees have oval light-colored blotches on the bark. Older trees are deeply furrowed and light to darker gray. **Leaves** are opposite, with 3–5 lobes, sparingly-toothed, 3–5 inches long, dark green above, pale green below. Sides of the terminal lobe are parallel or divergent; and notches between lobes are u-shaped. In autumn, leaves turn various shades of red, scarlet, orange or yellow.

Flowers are greenish-yellow, pendulous, appear on long, slender, hairy stalks and in clusters, with the leaves. The **fruit** is paired, round with wings that are about 1 inch long and slightly divergent. It ripens in the fall. The **twigs** are brown with sharp-pointed brown buds.

The **wood** is heavy, close-grained, strong and hard. It is used for furniture, flooring, tool handles, veneer, railroad ties, novelties, dowels, woodenware, canoe paddles, firewood and pulp. "Birds-eye" and curly-patterned maple is in high demand in the furniture and veneer industry. Historically, sugar maple was used to make parts for sleighs, sleds, pungs (low, one-horse box sleighs) and buggy shafts.

MAINE REGISTER OF BIG TREES 2008
Sugar Maple
Circumference: 213"
Height: 80' **Crown Spread:** 64'
Location: Palermo

Sugar maple buds are sharp-pointed and have scales that have a dark margin.

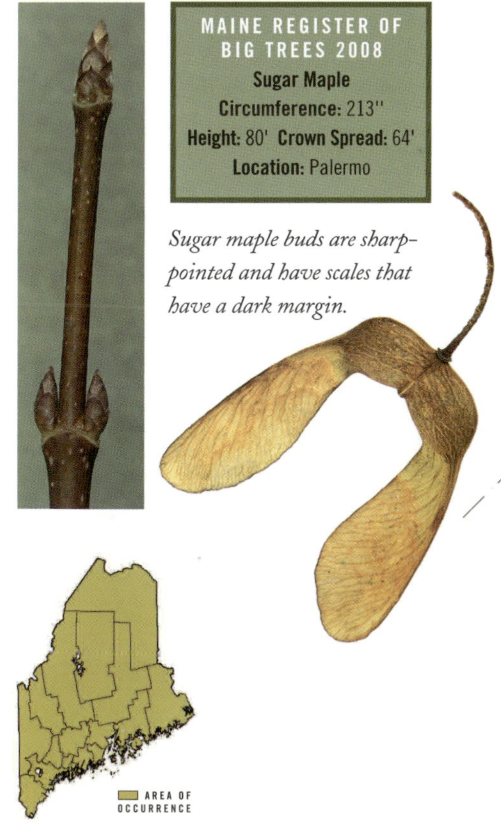

SILVER MAPLE *Acer saccharinum* L.

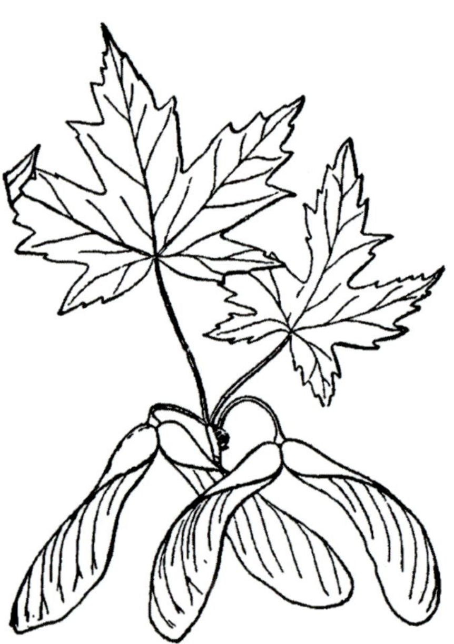

In Maine, silver maple is most common along major rivers.

Abundant in some localities, silver maple is a common tree, found throughout the state except along the coast. It grows largely on sandy banks along streams, usually attaining a height of 60–80 feet and a diameter of 2–3 feet. The trunk normally separates into 3 or 4 upright secondary stems, devoid of branches for some distance. The branches are long and slender, often pendulous.

The **bark** on young trees is smooth, gray, slightly tinged with red. On old trees, it is reddish-brown, furrowed, and separated into large thin scales that are loose at the bottom. Twigs are chestnut brown and shiny.

The **leaves** are opposite, deeply five-lobed; and the edges are irregular and sharply toothed. The upper surface is pale green, the lower, silvery white. They turn a pale yellow in fall.

The **flowers** are on very short stalks and in clusters. They are greenish-yellow or sometimes pinkish, opening early, long before the leaves appear.

The **fruit** is paired, winged and ripens in spring. Frequently, one of the pair does not fully develop. The **twigs** are curved upward at the tip, orange or red-brown above and green below, slender, with a bitter taste and a rank odor when broken.

The **wood** is softer than that of the hard maple, close-grained, not durable and easily worked. It is used to a limited extent for pulp.

Silver maple has large globe-shaped flower buds and smaller vegetative buds.

MAINE REGISTER OF BIG TREES 2008
Silver Maple Circumference: 316" Height: 89'
Crown Spread: 75' Location: Leeds

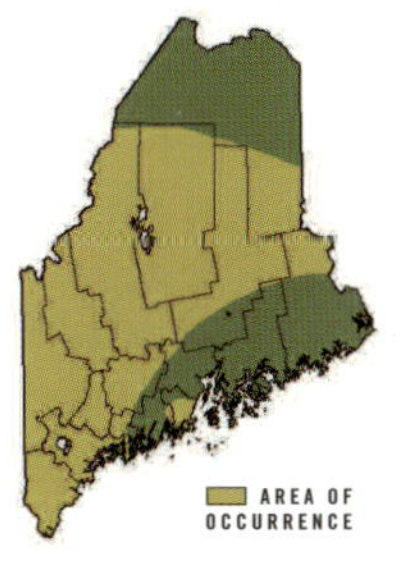

STRIPED MAPLE *Acer pensylvanicum* L.

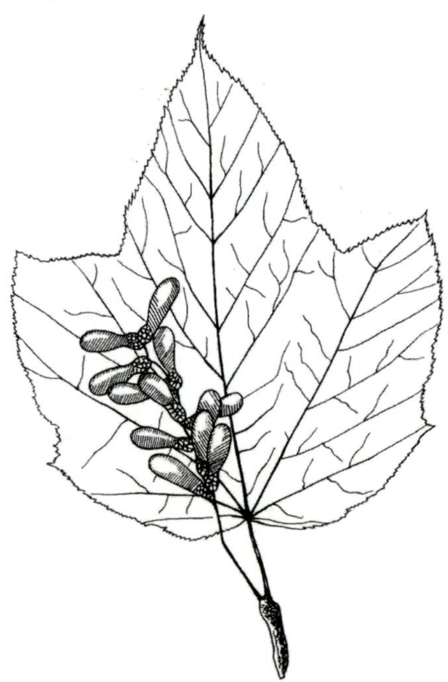

The striped maple is a shade-loving tree that is usually found growing with other hardwoods.

Striped maple or moosewood is common throughout the state. It is a shade-loving tree that is found growing with other hardwoods, or occasionally with conifers, on rich, moist soils or rocky slopes. Of little value except for its beauty, it rarely exceeds a height of 25 feet and a diameter of 8 inches. The branches are slender and upright, and the top narrow and often short.

The **bark** on the trunk is reddish-brown or dark green, and marked by whitish lines running lengthwise, which turn brown after a time. The **leaves** are three-lobed toward the apex, resembling a goose foot, opposite, finely toothed, pale green, 5–6 inches long and about as broad. In fall they turn light yellow.

The white and green-striped bark of the striped maple distinguishes it from any other native tree.

The **flowers** are bright yellow in slender drooping racemes that open the end of May or early June, when the leaves are fully grown. The **fruit** is paired, with wings moderately divergent, fully grown in late summer. It has a smooth, oval depression in the seed body. The **twigs** are smooth, reddish or greenish; the buds are valve-like, stout, stalked and without hairs.

The **wood** is close-grained, light and soft. During spring when the cambium layer is active, it is easy to make a whistle from the smaller branch sections.

MAINE REGISTER OF BIG TREES 2008
Striped Maple*
Circumference: 30"/31" **Height:** 45'/50'
Crown Spread: 21'/20'
Location: Harpswell /Monhegan Island *Tie

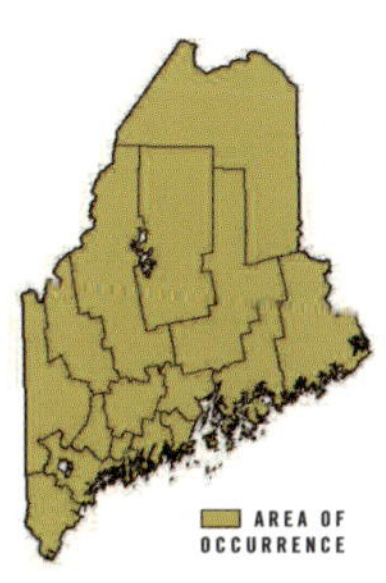

Mountain Maple *Acer spicatum* Lam.

Mountain maple occurs throughout Maine and is especially common in the northern part of the state. It grows as a small bushy tree, seldom over 30 feet in height. At times, the tree forms fairly dense thickets, due to its habit of growing in clumps. It grows best in a wet habitat or on damp, northern slopes. The slender twigs grow in a somewhat upright position.

The mountain maple grows as a small bushy tree, seldom over 30 feet in height.

M A P L E

Facing page, far left: Mountain maple leaves have deeply impressed veins on the upper surface.

The **bark** is reddish-brown to gray, thin and somewhat furrowed.

The **leaves** are opposite, three-lobed, shiny above, somewhat hairy below. They have rather coarse teeth and prominently sunken veins on the upper surface.

The **flowers** appear in June in long, hairy, yellow-green clusters after the leaves are full grown.

The **fruit** is paired, with wings slightly divergent, and occurs in ascending clusters. It has a wrinkled depression on the seed body and ripens in early fall.

The **twigs** are hairy, green, red or reddish-brown, not striped; and the pith is brown. The buds are hairy, valve-like, green, and only slightly stalked, slender and pointed.

The **wood** is close-grained, soft, light and not used commercially.

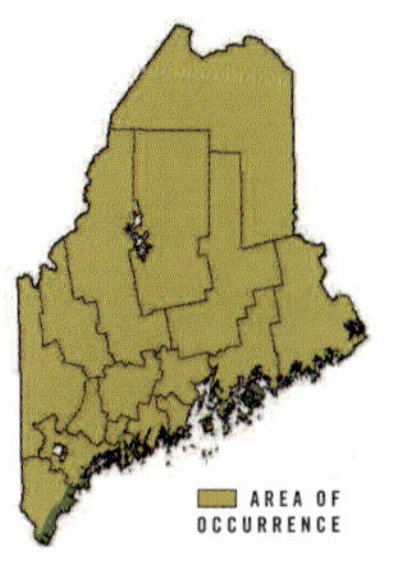

NORWAY MAPLE *Acer platanoides* L.

Native to continental Europe, Norway maple thrives in a wide variety of conditions, grows rapidly and casts a deep shade. Because of its aesthetic appeal and ease of propagation, it has been planted across Maine as a street and shade tree. It has escaped into the wild around many of our cities and towns, particularly in the southern half of the state. Because of its aggressive nature, Norway maple is considered to be a serious potential threat to our native flora and further planting of it is discouraged.

The **bark** of young trees is gray and smooth. Bark of older trees is gray-brown to almost black, and broken into long, interlacing vertical furrows.

Norway maple is not native to Maine. Because of its aggressive nature, it is considered to be a serious potential threat to our native flora and further planting of it is discouraged.

Norway maple drawing by Anna Anisko, used with the permission of the Pennsylvania Flora Project, Morris Arboretum of the University of Pennsylvania.

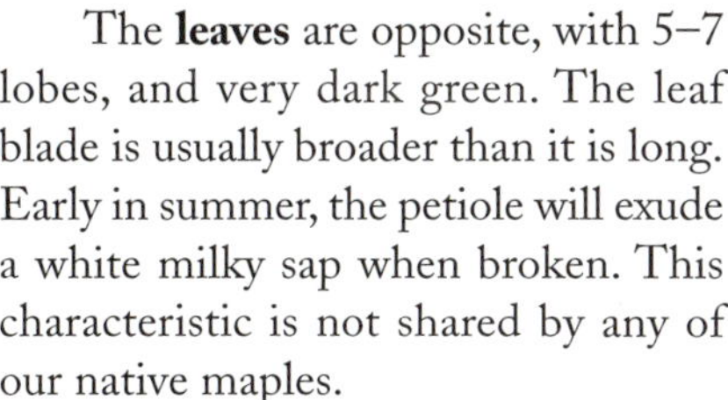

The **leaves** are opposite, with 5–7 lobes, and very dark green. The leaf blade is usually broader than it is long. Early in summer, the petiole will exude a white milky sap when broken. This characteristic is not shared by any of our native maples.

The **flowers** are bright yellow-green and appear in spring before the leaves.

The **fruits** are paired and diverge at a wide angle from each other. Each fruit has a leathery wing attached. The fruit and wing are about 2 inches long, flattened, light brown and mature in the fall.

The **twigs** are stout and brownish. The buds are large and green to purple with large bud scales.

The **wood** is occasionally used for firewood.

Norway maple fruit has a flattened seed body and a leathery, 2 inch wing.

MAINE REGISTER OF BIG TREES 2008
Norway Maple Circumference: 166'' **Height:** 70' **Crown Spread:** 86' **Location:** South Berwick

BOXELDER *Acer negundo* L.

Boxelder, or ashleaf maple, is apparently not native to Maine, but has been planted as an ornamental tree throughout the state and has escaped in localized areas near habitation. It was introduced along the St. John River in Aroostook County. It reaches a maximum height of 50 feet and diameter of about 2 feet in Maine. It is a short-lived, fast-growing, brittle tree, prone to wind and ice damage. It can become invasive.

Boxelder is a short–lived, fast-growing, brittle tree, prone to wind and ice damage. It can become invasive.

Boxelder twigs are purple and covered with a whitish waxy bloom.

The **bark** is light gray and smooth on young stems, becoming roughened and shallow-fissured on older trees.

The **leaves** are opposite, compound, usually 3–7 leaflets per leaf, rarely nine. The leaflets vary greatly in shape, often lobed and unlobed leaflets are found on the same leaf. Leaflets are occasionally divided into individual blades.

The **flowers** open just before the leaves in the spring and are yellow-green. They have no petals.

The **fruit** attains mature size in summer, ripening in autumn. It consists of a double-winged pair of seeds. Wings are only slightly divergent; and the seed body is wrinkled, three times longer than broad.

The **twigs** are smooth, rather stout, green or maroon, and covered with a white, chalky bloom. The bark yields a pungent odor when bruised.

The **wood** is light, soft, creamy white, often tinged with green, weak and close-grained. Occasionally it is used for pulp.

MAINE REGISTER OF BIG TREES 2008

Boxelder
Circumference: 115"
Height: 85'
Crown Spread: 66'
Location: Wilton

BIRCHES *The Important Distinctions*

	Paper Birch *Betula papyrifera*	**Gray Birch** *Betula populifolia*	**Yellow Birch** *Betula alleghaniensis*
BARK			
TEXTURE	Separates into thin, horizontal, papery layers	Does not separate into papery layers	Separates into thin, horizontal, ribbon-like strips
COLOR	Outer, chalky or grayish white; inner bark orange	Outer, chalky or grayish-white, dirty-looking; inner bark orange	Bright silvery gray or light yellow
ODOR	No odor	No odor	Wintergreen odor when young branches are scraped
LEAVES			
LENGTH	2–4 inches	2½–3 inches	3–4½ inches
OUTLINE	Egg-shaped	Triangular	Egg-shaped
MARGIN	Doubly toothed	Coarsely and doubly toothed	Coarsely and doubly toothed
SHAPE	Tip short pointed; base rounded	Tip long pointed; base truncated	Base unevenly rounded
SURFACE	Upper dark green, dull	Upper dark green and glossy	Upper dark green, dull and hairy
FLOWER			
STRUCTURE	3 catkins	Single or paired catkins	3–4 catkins
ARRANGEMENT	Clustered	Not clustered	Not clustered
BUDS			
TEXTURE	Sticky when squeezed	Not sticky	Smooth
SHAPE	Long, tapered	Short, globose	Long, sharp-pointed
SCALES	Without hairs	Without hairs	Hairy
COLOR	Reddish-brown	Red-brown to greenish-brown	Reddish-brown
TWIGS			
TEXTURE	Hairy, with spur shoots	Very fine, warty but not hairy, without spur shoots	Somewhat hairy, with spur shoots
COLOR	Depends on age	Dull gray or brown	Greenish or yellow-brown
ODOR	No wintergreen odor	No wintergreen odor	Slight wintergreen odor

	Sweet Birch *Betula lenta*	**Mountain Paper Birch** *Betula cordifolia*
BARK		
TEXTURE	Smooth on young trees; broken into irregular plates on older trees	Small portions peal away in thin sheets
COLOR	Dark to almost black	Whitish with a pink to salmon-colored tinged to reddish-brown
ODOR	Wintergreen odor when young branches are scraped	No odor
LEAVES		
LENGTH	3–5 inches	2–4 inches
OUTLINE	Egg-shaped	Egg-shaped
MARGIN	Singly and sharply toothed	Doubly toothed
SHAPE	Base heart-shaped	Tip short pointed; base heart-shaped
SURFACE	Upper dark green, dull; lower light yellow-green	Upper dull green
FLOWER		
STRUCTURE	3–4 catkins	2–4 catkins
ARRANGEMENT	Not clustered	Clustered
BUDS		
TEXTURE	Smooth	Sticky when squeezed
SHAPE	Long, sharp-pointed	Long, tapered
SCALES	Without hairs	Without hairs
COLOR	Chestnut brown	Brown
TWIGS		
TEXTURE	Smooth with spur shoots, no hairs	Sparsely hairy, often warty, with spur shoots
COLOR	Reddish-brown	Yellowish-brown to dark brown
ODOR	Strong wintergreen odor	No wintergreen odor

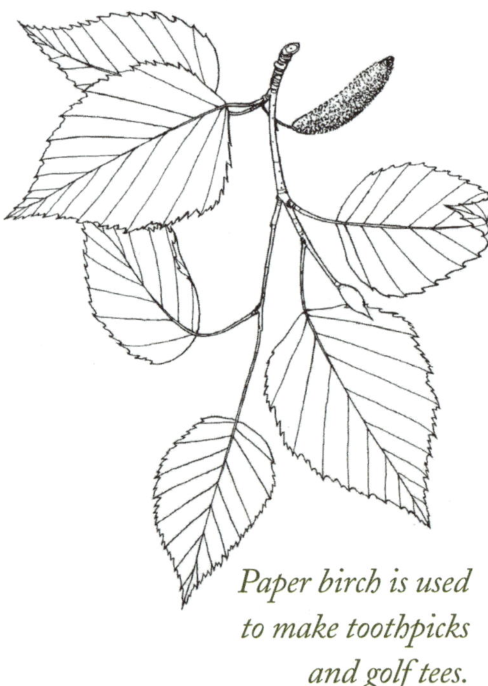

PAPER BIRCH *Betula papyrifera* Marsh.

Paper birch is used to make toothpicks and golf tees.

Paper, white or canoe birch is a common tree in all parts of the state; it occurs in pure stands or in mixture with other species. It reaches 60–70 feet in height and 1–2 feet in diameter. It grows along streams and on the borders of lakes and ponds, thriving best in a rich, moist soil.

When young, the branches are short, slender, spreading, somewhat drooping, and form a narrow, regular head. In the forest, the trunk is free from branches well up from the ground; and the tree forms an open, narrow and round-topped head.

The **bark** is a protective layer and should never be removed from living trees. On the trunk and limbs, it separates freely and easily into thin, papery sheets. The outer surface is white, the inner part bright orange. Seedlings or

very young trees have a darker colored bark, which gradually changes to a creamy-white.

The **leaves** are alternate, ovate, short-pointed, 2–4 inches long, thicker than those of gray birch, doubly-toothed, with the upper surface dark green and dull.

The **flowers** are in catkins. They open in early spring before the leaves. Those appearing in fall are dormant, staminate catkins and occur mostly in clusters of three.

The **twigs** are usually hairy and, unlike yellow birch, without a wintergreen taste. The buds are slightly sticky.

The **wood** is close-grained, moderately hard, and strong. It is used for woodenware, flatware and turned products including toys, dowels, furniture parts, pulp and firewood.

The tree gets the name of "paper birch" from how the bark was used by early settlers, and that of "canoe birch" because the bark was used to

Paper birch bark will peel off in large sheets, but it should never be removed from living trees.

make canoes. In the early spring paper birch sap contains considerable sugar. Historically paper birch was one of the most valuable tree species in Maine. In the past, the wood was used to make shoe pegs (used instead of nails in the manufacture of shoes) as well as a number of products that used to be made in Maine, but are now manufactured off-shore. These include clothespins, yarn spools, toothpicks, paper roll plugs and plywood.

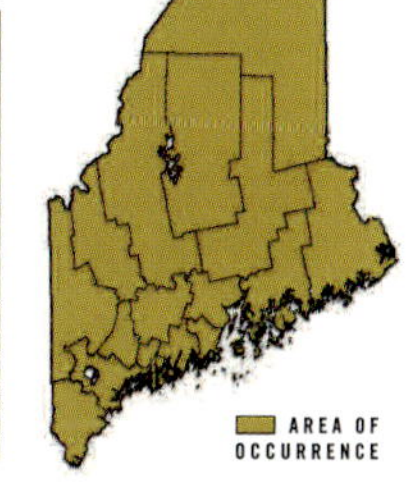

MAINE REGISTER OF BIG TREES 2008
Paper Birch
Circumference: 148"
Height: 72'
Crown Spread: 22'
Location: Alton

G R A Y   B I R C H *Betula populifolia* Marsh.

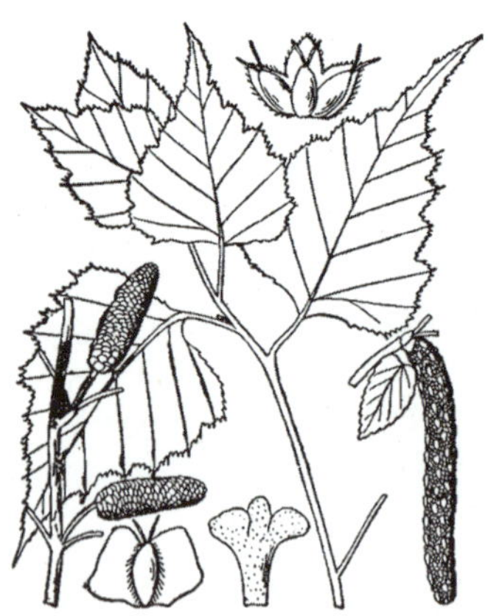

Gray birch is a short-lived and not particularly valuable tree. It occurs to some extent statewide, but is only abundant in the southern and eastern sections of the state. It is frequently found in old fields, burns and heavily-cut areas. This is a small tree that commonly reaches 20–30 feet in height and 4–8 inches in

Gray birch is a short-lived and not particularly valuable tree that is used primarily for pulp and firewood.

Gray birch has single or paired catkins in winter and spring.

diameter. It usually occurs in clumps and often leans. The branches are short, slender, frequently pendulous and contorted, and bend toward the ground when the tree is not crowded. The head is long, narrow, pointed and open.

The **bark** is close and firm, and does not easily separate into thin layers. The outer part is dull grayish-white or chalky. The inner portion is orange.

The **leaves** are 2½–3 inches in length, thin, long-pointed, triangular, alternate and doubly toothed. The upper surface is dark green and glossy. The slightest breeze causes them to flutter like those of the poplars, hence the scientific name *Betula populifolia* which means "birch with poplar leaves."

The **flowers** are produced in catkins. They open in early spring before the leaves. Those that appear in fall are male and usually solitary.

The **twigs** are the most slender of our native hardwoods. They are tough and wiry, dull gray or brown, hairless, and have a rough, warty surface. Dead twigs tend to stay attached to the trunk. This, plus the dirty appearance of the bark, makes this tree easy to recognize.

The **wood** is light, soft, often coarse-grained, and decays rapidly when exposed. It is occasionally used for pulp and firewood; in the past it was used for paper roll plugs.

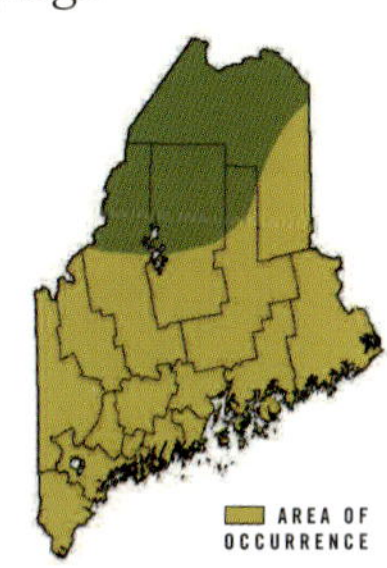

MAINE REGISTER OF BIG TREES 2008
Gray Birch
Circumference: 71"
Height: 65'
Crown Spread: 27'
Location: Richmond

YELLOW BIRCH *Betula alleghaniensis* Britt.

The yellow birch is one of our most valuable timber trees and makes excellent firewood.

Yellow birch is the largest of the native birches, growing to a diameter of 3 feet and a height of 70–85 feet. The spreading branches are somewhat pendulous; they form a broad, round-topped head in the open, but an irregular head in the woods. It grows well statewide on cool, moist sites, and is frequently mixed with beech and sugar maple, or with hemlock.

The **bark** on the branches and on the stems of young trees is very shiny, silvery-gray or yellowish-brown, separating into loose, thin, horizontal, often ribbon-like layers. On old trees, it is divided into large thin plates and is dull gray or black.

The **leaves** are 3–4½ inches long, ovate or nearly oblong, alternate; the edges are doubly toothed, the upper

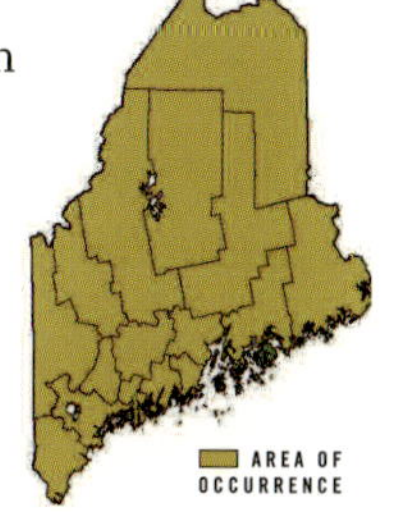

Yellow birch's bark peels in small curls. Very old trees have platy bark.

side dull, dark green and hairy. Leaves closely resemble those of eastern hophornbeam.

The **flowers** are in catkins. In winter there are 3–4 pre-formed staminate catkins on the shoots, but not in clusters. They open in the early spring.

The twigs are yellowish to dark brown and somewhat hairy. The young twigs are aromatic like sweet birch, although to a lesser degree. Both the buds and twigs have a pronounced wintergreen taste.

The **wood** is hard, strong, heavy and will take a good polish. It is close-grained and evenly textured. The heartwood, which makes up the bulk of the wood, has a pleasing reddish color; this is why it is sometimes called red birch. It takes stains easily, makes excellent veneer wood, and does not easily warp. It is also used for furniture, flooring, woodenware, lumber for interior finish, plywood, railroad ties, pallets, pulp, gunstocks and dowels. The yellow birch is one of our most valuable timber trees and makes excellent firewood. As with sweet birch, wintergreen oil was formerly distilled from twigs and branches.

> **MAINE REGISTER OF BIG TREES 2008**
> **Yellow Birch Circumference:** 200" **Height:** 48' **Crown Spread:** 91'
> **Location:** Deer Isle

S WEET B IRCH *Betula lenta* L.

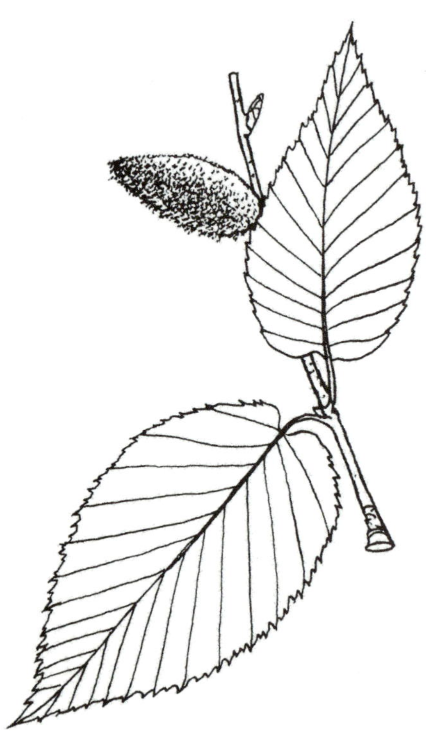

Sweet, black or cherry birch is found, though uncommonly, in the southern third of the state. It inhabits the banks of streams or moist, rich upland soil. It is a handsome tree with a tall dark stem, and spreading, slender, horizontal branches that are pendulous at the ends. It has a graceful, open, narrow head, which in full sun becomes round and symmetrical. It grows to a height of 60–70 feet and a diameter of 1–2 feet.

The **bark** on the trunk of old trees is dark to almost black, and separates into large, thick, irregular plates. On young trees and branches, it is smooth, shiny, dark brown tinged with red, aromatic, and has a very pronounced wintergreen flavor.

The **leaves** are alternate, 3–5 inches long, aromatic, ovate or somewhat

The name "cherry birch" is applied to this tree because of the resemblance of the bark on old trunks to that of the black cherry.

oblong, and sharply toothed. The upper surface is dark green and dull; the lower surface is light yellow-green.

The **flowers** are produced in catkins. The winter shoots support 3 to 4 staminate catkins. They open just before the leaves unfold in the spring.

The **wood** is hard, heavy, strong and can be beautifully polished. It is prized for use in the manufacture of furniture and it makes excellent firewood. Limited amounts are used as pulpwood. Historically, oil with some medicinal value was obtained from the branches and bark by distillation, and was generally known as wintergreen oil.

The name "cherry birch" is applied to this tree because of the resemblance of the bark on old trunks to that of the black cherry.

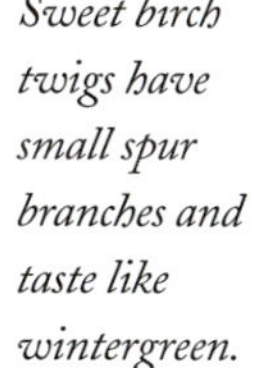

Sweet birch twigs have small spur branches and taste like wintergreen.

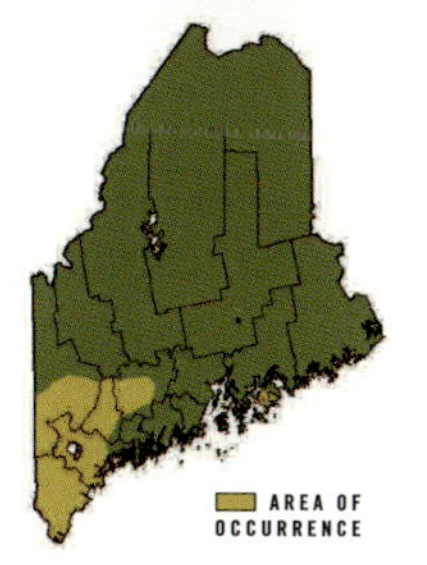

MAINE REGISTER OF BIG TREES 2008
Sweet Birch Circumference: 139" Height: 82'
Crown Spread: 48' Location: Gorham

MOUNTAIN PAPER BIRCH
Betula cordifolia Regel.

Mountain paper birch is closely related to paper birch, and has been designated as a variety of the species by some authors (*Betula papyifera var. cordifolia* (Regel) Fern.). It is known at many points in Maine, particularly on mountain slopes, coastal headlands and islands east of Mount Desert Island. It often grows as a clump of several stems. It can grow to about 60 feet in height and 1 foot or more in diameter.

The **bark** of young trees and branches is dark reddish-brown and does not peel. The bark of older trees will separate into thin, papery layers. In mature trees, bark color ranges from whitish with a pinkish or salmon-colored tinge to reddish-brown or bronze.

Mountain paper birch is most often found along the coast and at high elevations.

The **leaves** are egg-shaped with heart-shaped bases, abruptly pointed, and coarsely doubly-toothed. The flowers are borne in catkins. Dormant male catkins in clusters of 2–4 are visible during winter. Both male and female catkins expand in spring.

The **fruit** is a tiny nutlet with 2 small lateral wings. It matures in fall and is dispersed during the fall and winter. Large numbers of birch seed can often be seen on the surface of the snow.

The **wood** is similar to that of paper birch; and the two species are usually not separated. It is used for turnery products, cabinetry, pulp and fuel.

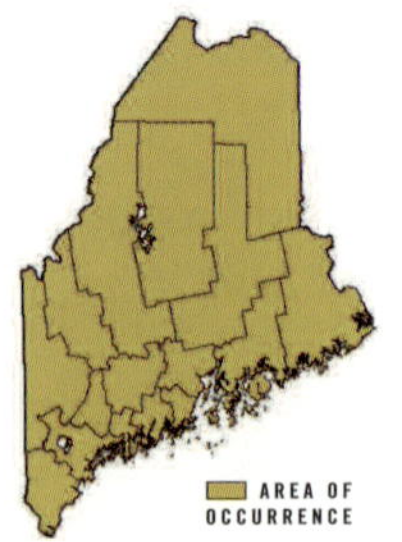

Maine's trees have been used to make hundreds of different products, many of which are now obsolete, including elm hub stock, shown here in an early photo from Cumberland, Maine.

Eastern Hophornbeam
Ostrya virginiana (P. Mill.) K. Koch

Eastern hophornbeam or iron-wood is a small tree with either an open or rounded crown. It reaches a height of 20–30 feet and a diameter of 6–10 inches. The branches are long and slender, with ends that are somewhat drooping.

It is a fairly rapid grower, especially in good soil. It grows on slopes and ridges having a dry, gravelly soil, and is often found in the shade of other species.

The **bark** is gray, and separates easily into thin, narrow, vertical scales, becoming finer and stringy on older trees.

The **leaves** are 2–3 inches long, egg-shaped to nearly oblong in outline, widest in the middle, hairy on both surfaces, alternate and sharply toothed. They are somewhat like those of yellow birch.

The **flowers** occur in catkins, which open with the leaf buds. The male catkins are pre-formed in the fall and are usually in clusters of three.

The **fruit** is bladder-like, encloses a ribbed nutlet and occurs in clusters.

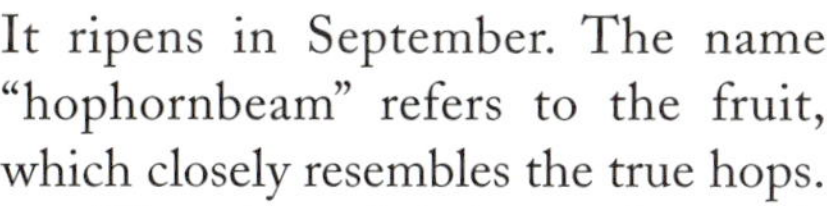

It ripens in September. The name "hophornbeam" refers to the fruit, which closely resembles the true hops.

The **twigs** are light brown, fine, tough and wiry, and have a small green pith.

The **wood** is very close-grained, heavy, very strong and is exceedingly hard when seasoned. It is used for tool handles, wedges for directional felling of trees, and firewood. In the past, it was used to make trip stakes on log hauling trucks (which contain, then release the logs from the truck beds), wagon tongues (the shaft where the horses are hitched) and other parts.

MAINE REGISTER OF BIG TREES 2008
Eastern Hophornbeam*
Circumference: 77"/70" **Height:** 63'/67'
Crown Spread: 38'/42'
Location: Livermore Falls/Pownall
*Tie

Eastern hophornbeam buds have tiny vertical grooves that can be seen with a magnifying glass.

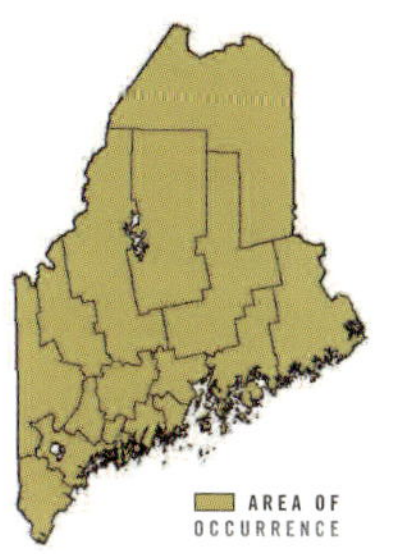

American Hornbeam
Carpinus caroliniana Walt.

American hornbeam, blue-beech or musclewood occurs west of the Penobscot River in the southern third of the state. The American hornbeam is most commonly found inhabiting wet woods and the borders of swamps and streams. It is a small, slow-growing tree 10–25 feet tall and 4–10 inches in diameter. The branches are crooked. The trunk is characteristically ridged, or fluted longitudinally.

The **bark** is smooth and grayish-blue. The leaves are alternate, egg-shaped or oval, 2–3 inches long, sharply toothed, smooth above and hairy below. They turn a brilliant scarlet in autumn.

Close-grained, strong, tough and durable, American hornbeam wood is used for levers, handles and wedges.

American hornbeam bark has a sinewed appearance that gives it its alternate name, "musclewood."

The **flowers** are produced in catkins that open in spring before the leaves.

The **fruit** is a ribbed nutlet, which is attached to the base of a three-lobed bract, and is borne in open clusters.

The **twigs** are reddish-brown, slender and tough. Buds are also reddish-brown and slender, and sharp-pointed.

The **wood** is close-grained, compact, strong, tough and durable. It is used for levers, handles, and wedges.

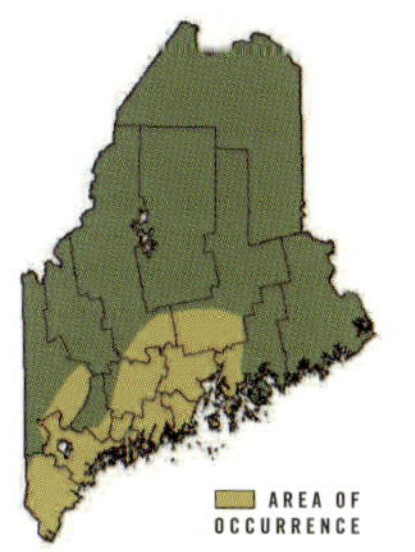

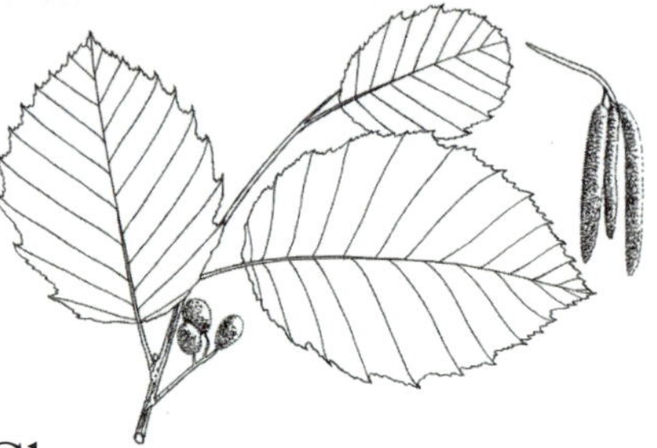

SPECKLED ALDER
Alnus incana ssp. Rugosa (Du Roi) Clausen

Speckled alder is very common in Maine, usually growing in wet areas along brooks, in swamps and in pastures. It sprouts readily and is a nuisance on pasture land. Alder usually occurs as a shrub, rarely as a small tree. It is seldom more than 4 inches in diameter and 20 feet in height.

The **bark** is smooth, dark chocolate brown, and marked with white, horizontal, elongated spots called lenticels.

The **leaves** are alternate, 2–3 inches long, usually broadly ovate; and the texture is rough or rugose as the scientific name implies. The edges are unevenly or doubly-toothed.

The **flowers** are in catkins, and open before the leaves in spring. The purplish, wax-like male catkins are pre-formed the previous fall. The fruit is woody and cone-like, with a very short stalk.

The **winter buds** are short-stalked and maroon, with few scales showing.

The **twigs** are reddish-brown; the pith is triangular in cross section.

The **wood** is light and soft, and has very little commercial use. The wood discolors very rapidly on exposure to air. Baskets for the florist industry are made from small diameter stems. In the past, the wood was used in hand forges, because of the intense heat it produces when burned.

Two other species, **green or mountain alder**—*Alnus viridis* (Vill.) Lam. & DC. *Spp. Crispa* (Ait.) Tirrill)—and **hazel alder** (*Alnus serrulata* (Ait.) Willd.) occur as shrubs.

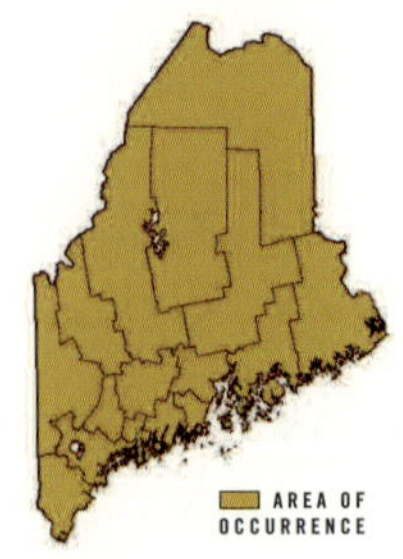

Mechanical equipment, including the Lombard log hauler, began to replace horses in the Maine woods in the early 1900s.

AMERICAN BEECH *Fagus grandifolia* Ehrh.

American beech occurs statewide, growing up to 70 feet in height and 1–3 feet in diameter. Although it grows best on rich upland soil, beech is common and sometimes forms nearly pure stands, with shoots often springing up from the trees' roots. Beech bark disease causes significant mortality in Maine. The disease results when bark, attacked and altered by the beech scale insect, *Cryptococcus fagisuga* is invaded and killed by fungi, primarily *Nectria coccinea* var. *faginata*.

The **bark** is light gray and smooth unless affected by beech bark disease. The bark of trees affected by the disease is rough and pockmarked with small cankers.

The **leaves** are alternate, 3–5 inches long, elliptic, acutely pointed, with coarse and hooked teeth. The margin between the teeth is nearly straight. Dead leaves are light tan and tend to remain on trees into winter.

The **fruit** consists of a bur, which usually contains 2 triangular edible nuts. These nuts are sweet and are an important food source for wildlife. Trees that bears have climbed to eat beech nuts show claw marks on the bark. The **winter buds** are long, slender, many scaled, and sharp-pointed.

The **wood** is strong, hard and tough, but not durable. Current uses include for pulp, pallet stock and firewood. In the past, it was used for clothespins, furniture, handles, woodenware, railroad ties, dowels and flooring.

European beech (*Fagus sylvatica* L.), Purple (*Fagus sylvatica var. atropunicea* Weston) and Copper beech (*Fagus sylvatica var. cuprea* L.) are species of European origin planted in southern and central Maine as ornamentals.

Smooth barked beech (left) have become a rarity in Maine. The bark of most trees (right) is roughened by cankers caused by beech bark disease.

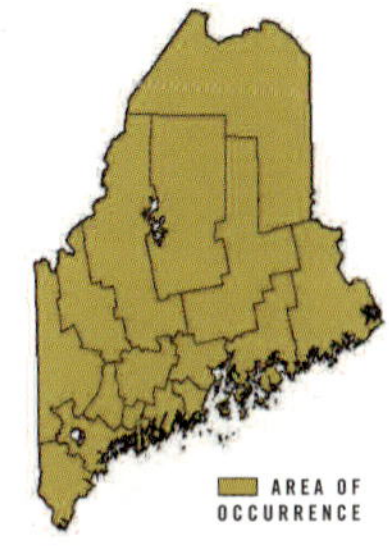

BLACK OAKS *The Important Distinctions*

	Northern Red Oak *Quercus rubra*	**Black Oak** *Quercus velutina*	**Scarlet Oak** *Quercus coccinea*	**Bear Oak** *Quercus ilicifolia*
Black oak group: leaves sharp-tipped, acorns mature in two years and are hairy inside.				
BARK				
TEXTURE	Slightly ridged	Deeply fissured blocky, ridges very dark	Ridges small and irregular	Smooth, with a few raised lenticels
COLOR	Dark gray to black; inner reddish	Black; inner—deep orange or bright yellow	Dark gray to black; inner—pale red or gray	Gray to dark brown
LEAVES				
LENGTH	5–8 inches	5–6 inches	3–6 inches	3–6 inches
COLOR	Surface—dull, dark green; below—yellow-green	Surface—dark green and shiny	Surface—bright green and shiny; below—paler	Surface—dark green; below—white or gray
SURFACE	Smooth beneath	Hairy beneath	Smooth beneath	Hairy beneath
ACORNS				
SIZE	2–4 times longer than cup	Twice as long as cup	Twice as long as cup	Small, nearly hemispherical, striped above middle
CUP	Saucer-like, with scales fused	Bowl-like, with dull scales	Bowl-like, with shiny scales	Shallow
BUDS				
SHAPE	Conical, smooth	Large, decidedly angled	Small, rounded	Small, short, blunt-pointed
COLOR	Chestnut brown	Yellowish-gray	Light brown	Reddish-brown
SCALES	Silky at tip	Coated with matted wool-like hairs	Hairy at tip only	Loose scales

WHITE OAKS *The Important Distinctions*

	White Oak *Quercus alba*	**Chestnut Oak** *Quercus prinus*	**Bur Oak** *Quercus macrocarpa*	**Swamp White Oak** *Quercus bicolor*
White oak group: leaf lobes rounded, acorns mature in one year and lack hairs inside.				
BARK				
TEXTURE	Ridges broad, flat, flaky	Deeply furrowed	Deeply furrowed, flaky	Deeply fissured; broad, flat ridges, flaky
COLOR	Light gray	Reddish-brown to dark brown	Grayish	Grayish-brown, inner bark orange
LEAVES				
LENGTH	4–7 inches	4–8 inches	6-12 inches	4–6 inches
COLOR	Surface—bright green	Surface—yellow green	Surface—dark green	Surface—dark green
SURFACE	Upper—dull	Upper—shiny	Upper—shiny	Upper—dull
SHAPE	Lobes rounded; cleft to midrib	Narrowly elliptical; shallow rounded lobes	Violin-shaped; lobes rounded	Slightly lobed
ACORNS				
SIZE	Very fine	Three times as long as cup; long and wrinkled	Twice as long as cup	Three times as long as cup
CUP	Short-stalked	Hairy, moderately long-stalked	Margin fringed with long, hair-like scales, short stalked	Margins fringed with scales, long-stalked
BUDS				
SHAPE	2–4 times longer than cup	Droadly ovoid, sharp-pointed	Broad ovoid, blunt or sharp-pointed	Roundish, blunt-pointed
COLOR	Dark red-brown	Yellowish-brown	Reddish-brown	Brown
SCALES	Without hairs	Without hairs	Coated with soft hairs	Small, without hairs

NORTHERN RED OAK *Quercus rubra* L.

Northern red oak is the most common species of oak in Maine.

Northern red oak is the most common oak species in Maine. It occurs state-wide but is most abundant in the southern half of the state. Best growth is attained on rich upland soils. It grows to a height of 60–80 feet and a diameter of 2–3 feet, forming either a narrow or broad head. The branches are stout, horizontal or upright.

The **bark** on the trunks of old trees is dark gray or nearly black, and is divided into rounded ridges. On younger trees and branches, it is smooth and greenish-brown or gray. The inner bark is reddish. The **leaves** vary in shape, are 5–8 inches long, alternate, are dull, dark green above

and yellow-green below, and have bristle-tipped lobes. Some dead leaves may remained attached in winter.

The **flowers** appear in May, when the leaves are only partly grown. The **fruit** ripens the second year. The acorn is broad, large, 1–1¼ inches long, and up to four times longer than the shallow cup. Red oak acorns are a major source of food for many species of wildlife. The inside lining of the acorn is densely hairy. Its tannic acid content makes it bitter. The **twigs** are smooth, greenish to reddish-brown, and have a star-shaped pith.

The **wood** is hard, strong and relatively heavy. It is used for furniture, interior finish, planks and frames, lobster trap runners, flooring, piling, cross-ties, timbers, pallets, dowels and firewood. Historically, it was used for shipbuilding (ribs, beams and timbers), weir poles (some 60 feet long), as kiln wood and fence posts (when split).

Northern red oak often has a reddish coloration in the bark fissures.

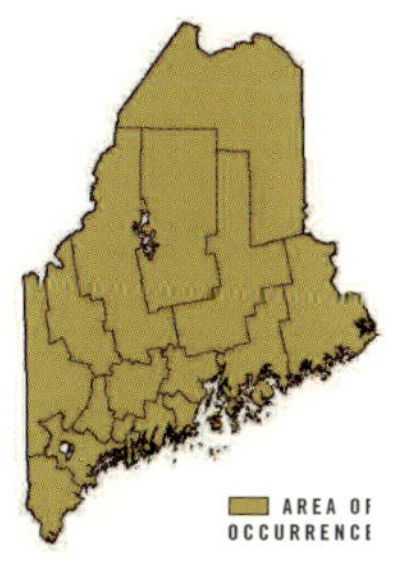

BLACK OAK *Quercus velutina* Lam.

B lack or yellow oak is found in southern Maine from Lincoln and southern Oxford counties southward; it is common near Fryeburg. It grows on dry ridges and gravel uplands. The branches are slender; and the head is narrow and open. It grows to a height of 50–60 feet and a diameter of 1–2 feet.

Black oak is used to a limited extent for interior finish, shipbuilding, flooring, piling, cross-ties, timbers, pallets, dowels, and firewood.

Black oak buds and twigs are covered with tan fine hairs.

The **bark** is smooth and dark gray or brown on young stems. On old trees, it is divided by deep fissures into broad, rounded ridges and is dark, almost black. It is rougher than red oak. The inner bark, which is characteristically bright orange or bright yellow, was used in tanning.

The **leaves** are alternate, 5–6 inches long, varying much in shape and general outline, usually seven-lobed with bristle points. The upper surface is glossy and dark green; the under surface is generally hairy with more obvious, rusty hairs in axils of veins.

The **flowers** appear in May when the leaves are only partly grown.

The **fruit**, which is bitter, matures the second season. The acorn is ½–¾ inch long, almost twice as long as the cup and one-half enclosed by it.

The **twigs** are smooth; buds are densely hairy, angled and yellowish-gray.

The **wood** is hard, heavy, strong and coarse-grained. It is used for the same purposes as scarlet oak

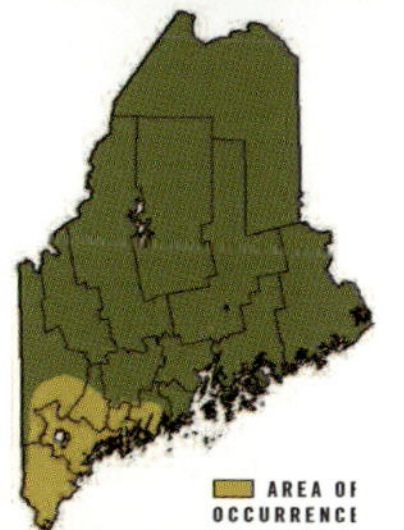

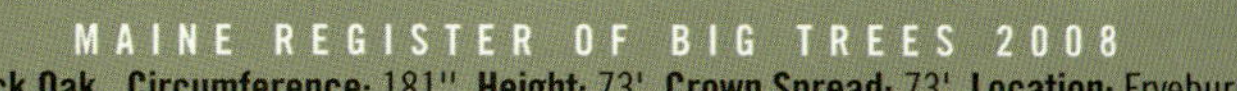

SCARLET OAK *Quercus coccinea* Muenchh.

Scarlet oak occurs rarely in the southern tip of Maine. It grows on the dry soil of ridges and uplands in York, Cumberland and Androscoggin counties.

In Maine, scarlet oak is a smaller tree than red oak, growing to a height of 30–50 feet and 1–2 feet in diameter. The branches are slender and form an open, narrow head.

In Maine, scarlet oak is a smaller tree than red oak.

The **bark** on the trunks of old trees is separated into irregular ridges by shallow fissures, and is dark gray with a reddish inner bark.

The **leaves** are alternate, 3–6 inches long with a variable outline. The upper surface is bright green and shiny; the lower is paler and less shiny. Lobes are sharp-tipped. In fall, the leaves turn a deep scarlet, which accounts for the common name of the tree.

The **flowers** appear in May when the leaves are only partly developed.

The **fruit** ripens the second year. The acorn is about ½ inch long, about twice as long as the cup, and is from one-third to one-half enclosed by the cup. It is quite bitter.

The **wood** is hard, strong and heavy, but coarse-grained. It is used to a limited extent for interior finish, shipbuilding, planks and frames, flooring, piling, cross-ties, timbers, pallets, dowels, and firewood.

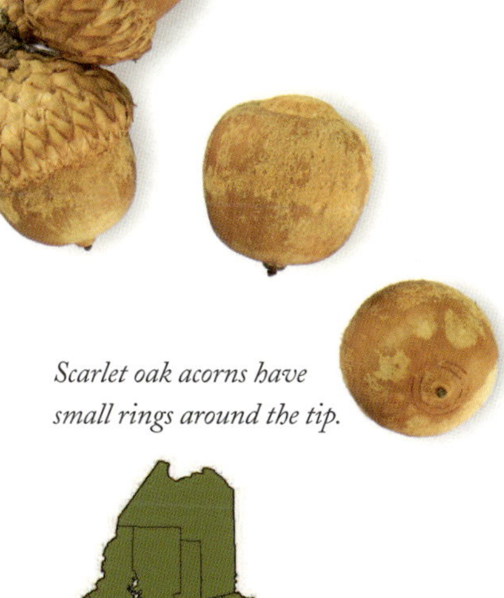

Scarlet oak acorns have small rings around the tip.

MAINE REGISTER OF BIG TREES 2008
Scarlet Oak Circumference: 73" **Height:** 73' **Crown Spread:** 40' **Location:** Yarmouth

BEAR OAK *Quercus ilicifolia* Wangenh.

Bear or scrub oak is a small, thicket-forming, shrubby tree usually less than 20 feet high. It is common on the sand barrens of southern Maine, extending into Oxford County where it is common on the barrens surrounding the Saco River near Fryeburg. It also occurs in eastern Hancock County, and on rocky ridges and barren ledge sites along the coast.

The **bark** is smooth, gray-brown, and has a few raised lenticels. On larger trees the bark may become rough and scaly.

Bear oak wood is occasionally used as fuel, but is generally not considered to be of commercial value.

Bear oak twigs are finer than the other native oaks and covered with hairs.

The **leaves** are the primary distinguishing feature. They have 5–9 bristle-tipped lobes, are whitened on the underside, alternate and 2–4 inches long. The second set of lobes from the base tends to be much larger than others.

Male **flowers** are pale reddish-green catkins that appear in May when the leaves are only partially grown. The **fruit** is a dark brown acorn 2/5–4/5 inch long that matures in the fall of the second season.

The **twigs** are slender and densely hairy during the first year. The terminal buds are clustered, chestnut brown and blunt-pointed. The lateral buds are of similar size to the terminal.

The **wood** is occasionally used as fuel, but is generally not considered to be of commercial value.

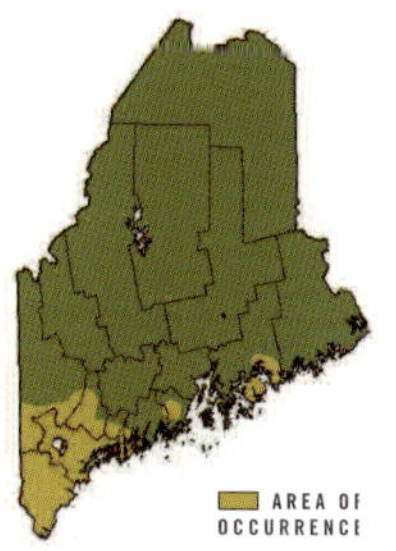

W H I T E O A K *Quercus alba* L.

White oak, which gets its name from the color of the bark, occurs naturally in southern and central Maine as far north as Oakland in northern Kennebec County. It grows on sandy land, gravelly ridges and moist bottomland, but makes the best growth on rich, heavy, upland soils. In good situations, it attains a height of 60–70 feet and a diameter of 3–4 feet. When not crowded by other trees, the

White oak wood is strong, heavy, hard and durable, making it ideal for use in flooring, furniture and boatbuilding.

White oak acorns mature in one year. They contain less tannin than red oak acorns and are preferred by wildlife.

bole (trunk) is short, the limbs are large and diverging, and the head is broad and rounded. In the forest, it has a long bole and a narrow head.

The **bark** on the trunk is separated into thin, irregular flakes and varies from light to ashy-gray.

The **leaves** are usually nine-lobed. The lobes are rounded, slightly cleft or cleft nearly to the midrib, alternate, 4–7 inches long, bright green above, pale green or whitish beneath. They sometimes remain on the tree during winter.

The **flowers** come out in May when the leaves are half grown.

The **fruit** ripens in September of the first year. The acorn is about ¾ inch long, 2–4 times longer than the cup, and about one-quarter enclosed by it. The fruit is edible. American Indians pounded it into a flour and bleached out the tannin with hot water.

The **twigs** are gray to purple; buds are blunt-pointed, and scales are without hairs.

The **wood** is strong, heavy, hard and durable. It is used for ship and boatbuilding, railroad ties, piling, agricultural implements, interior finish, furniture, flooring, pulp, and firewood. In the past, it was used for deck planking on ships, tight cooperage (tight casks capable of holding liquid such as whiskey), and spokes and rims of wooden wheels.

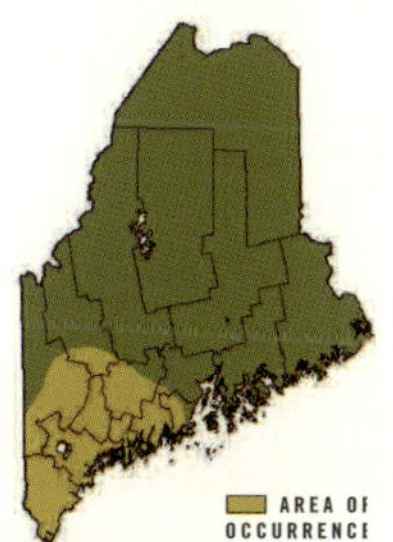

M A I N E   R E G I S T E R   O F   B I G   T R E E S   2 0 0 8

White Oak   Circumference: 239"   **Height:** 64'   **Crown Spread:** 95'   **Location:** Pittston

CHESTNUT OAK *Quercus prinus* L.

Chestnut oak only occurs in the southern tip of Maine. It is found on Mt. Agamenticus in the town of York and has been reported from Oxford County. In Maine, trees grow 12 inches or more in diameter and about 40 feet in height.

The gray-brown **bark** is smooth on young trees, but becomes thick and very deeply furrowed on older trees. The **leaves** are similar to those of the American chestnut. They are yellow-

Due to chestnut oak's rarity in Maine, it is not used commercially here.

*They gray bark of the chestnut oak
is very deeply furrowed.*

green above, hairy below, narrowly elliptical with shallow rounded lobes without bristle tips. They are often widest above the middle.

Male **flowers** are yellow-green, borne on catkins and appear in May. Female flowers are reddish, borne in spikes with the leaves in mid-spring. The edible **fruit** is a large, 1½ inch long, ellipsoid acorn that matures in one season. Its cup encloses about half of the acorn. The **twigs** are hairless and orange-brown to gray. The chestnut brown buds are clustered toward the end of the twig, pointed, and quite long and narrow in shape.

The **wood** is similar in character to white oak and has similar uses. In areas where chestnut oak is more abundant, it is sold as white oak. Due to its rarity in Maine, it is not used commercially here.

MAINE REGISTER OF BIG TREES 2008

Chestnut Oak   Circumference: 191''   **Height:** 90'   **Crown Spread:** 72'   **Location:** Yarmouth

BUR OAK *Quercus macrocarpa* Michx.

Bur oak is found in the southern two-thirds of the state and is locally plentiful in central Maine. It is quite common along the Sebasticook River, the lower Penobscot basin, and east into Hancock County. It grows in low, rich bottomland, and is rarely found on dry soil. It has a broad top of wide, spreading branches. The trunk is often clear of limbs for two-thirds or more of its length. It attains a height of 60–70 feet, and a diameter of 2–3 feet.

Very durable, hard, heavy and strong, bur oak is used for for cabinetry, barrels, hardwood flooring and fence posts.

Bur oak twigs have ridges of corky bark, a characteristic that is not shared by any of our other native oaks.

The **bark** is grayish, deeply furrowed and broken into plate-like irregular scales.

The **leaves** are roughly violin-shaped in outline, with rounded lobes that are not generally as deeply cut as the white oak. The upper end of the leaf is widest. They are alternate, dark green and shiny on the upper surface, pale green or silvery-white on the lower.

The **flowers** appear in May when the leaves are partly formed.

The **fruit**, which is edible, matures the first year and is usually solitary. It varies in size and shape. The acorn is about ¾ inch long, and about half enclosed by the cup. The margin of the cup is fringed with long, hair-like scales. The **twigs** have corky wings or ridges.

The **wood** is very durable, hard, heavy and strong. It is used for the same purposes as white oak.

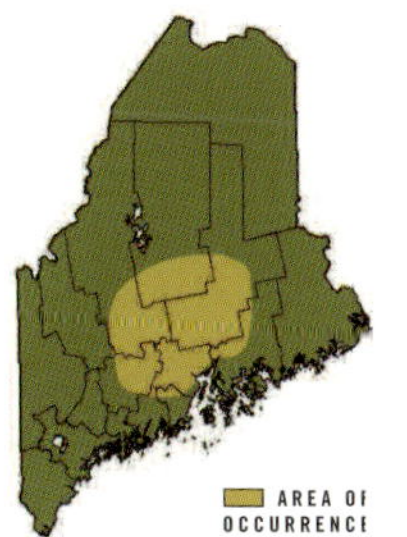

Swamp White Oak *Quercus bicolor* Willd.

Swamp white oak is not abundant, but occurs in small, widely scattered groves in York and Androscoggin counties. It grows in moist, fertile soil on the borders of swamps and along streams.

Swamp white oak grows to a height of about 50 feet and a diameter of 2–3 feet. The limbs are small, usually pendulous. The head is narrow, open and round-topped.

The **bark** on old trees is deeply furrowed, divided into broad, flat ridges, flaky and grayish-brown. On

Swamp white oak is not abundant, but occurs in small, widely scattered groves.

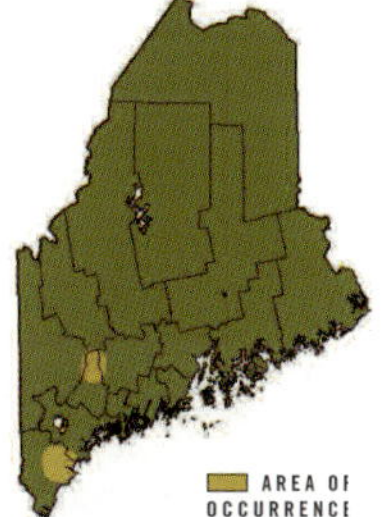

AREA OF
OCCURRENCE

young trees and branches, it is smooth
and separates into papery scales that
hang loosely. The inner bark is orange.

The **leaves** are alternate, 4–6 inches long and slightly lobed. The upper
surface is dark green and shiny; the
lower, pale white or tawny.

The **flowers** appear in May when
the leaves are not more than half-grown.

The **fruit** matures the first season.
The acorn has a long stalk, is about 1
inch long, three times as long as the cup
and about one-third enclosed by it.

The **twigs** have a yellowish or a
light orange to reddish-brown bark.

The **wood** is strong, heavy, hard
and used for the same purposes as the
white oak.

*Swamp white oak twigs have small rounded
buds. The bark of the twigs and small branches
tends to peel and flake.*

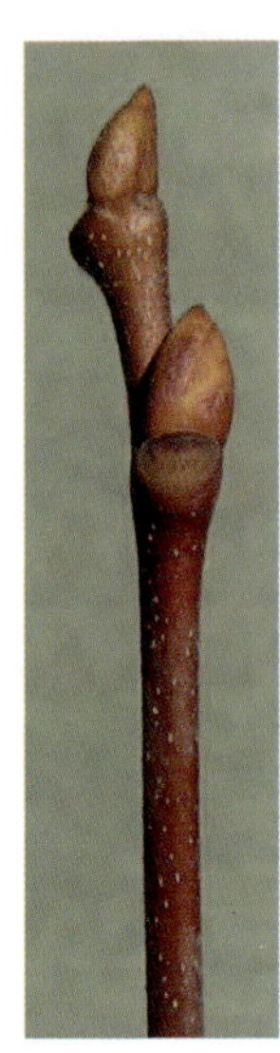

American Chestnut
Castanea dentata (Marsh.) Borkh.

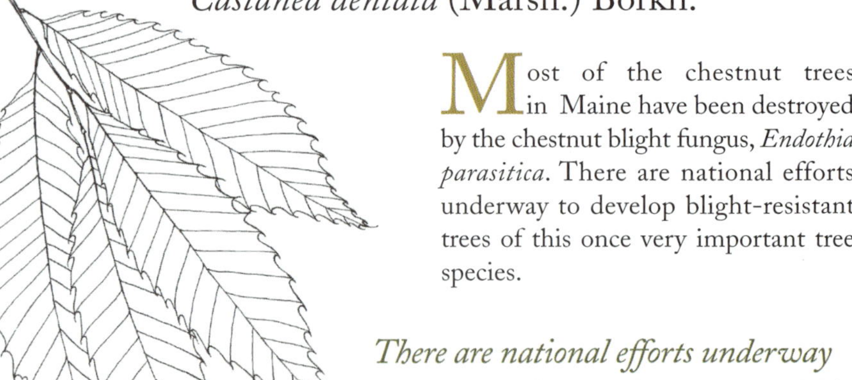

Most of the chestnut trees in Maine have been destroyed by the chestnut blight fungus, *Endothia parasitica*. There are national efforts underway to develop blight-resistant trees of this once very important tree species.

There are national efforts underway to develop blight-resistant trees of this once very important tree species.

The fruit of American chestnut is contained in a very prickly bur.

The natural range of American chestnut only extended into southern and central portions of the state. Chestnut now occurs infrequently, usually as sprout growth, in the southern half of the state on rich, well-drained soil. It has been planted occasionally as far north as Orono. The tree grows rapidly. In the forest, it has a tall, straight trunk free of limbs, and a small head. When not crowded, the trunk divides into 3 or 4 limbs and forms a low, broad top. It reaches a height of 60–70 feet and a diameter of 15–30 inches.

The **bark** on the trunks of old trees is dark brown and divided into broad, flat ridges by shallow, irregular fissures. On young stems, it is smooth and dark gray with a green tinge.

The **leaves** are coarsely-toothed and hooked, with the leaf margin rounded between the teeth. Leaves are alternate, 5–8 inches long, yellow-green and smooth on both surfaces.

The **fruit** is a prickly bur containing 2–3 nuts tipped with hairs. The inner lining of the bur is plush-like.

The **nuts** contain a sweet meat; they were once gathered in large quantities for the market.

The **wood** is soft, very durable, strong and splits easily. It is used for interior finishing and was once in much demand—prior to the blight—for telephone poles, railroad ties, beams and timbers up to 50 feet in length, furniture stock and fence posts. The durability of the wood is due to the tannic acid that it contains.

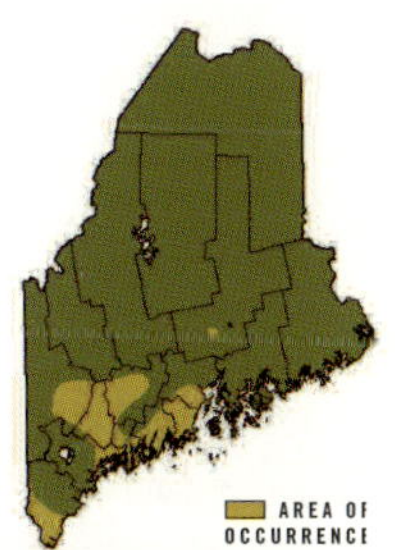

MAINE REGISTER OF BIG TREES 2008
American Chestnut Circumference: 117″ **Height:** 45′ **Crown Spread:** 44′ **Location:** Orono

To keep logs sorted as they were transported to mills, each company chopped a brand into its logs. A Great Northern Paper Company brand can be seen in the foreground.

ASHES *The Important Distinctions*

	White Ash *Fraxinus americana*	**Green Ash*** *Fraxinus pennsylvanica*	**Black Ash** *Fraxinus nigra*
LEAVES			
LEAFLETS	5–9, usually 7	7–9	7–11
DESCRIPTION	Leaflets are mostly entire, borne on stalks, without hairs below. Turn purple in autumn	Leaflets borne on stalks. Hairy below and on rachis. Turn yellow or bronze in autumn.	Toothed leaflets which are without stalks except the one at the end. Hairs lacking below except for buff-colored hairs at the junction of the leaflets and the rachis. Turn yellow in autumn.
BUDS			
SIZE	⅛ inch	⅛ inch	Less than ¼ inch
SHAPE	Blunt-pointed	Cone-shaped	Sharply-pointed
COLOR	Brown	Brown with rusty or dull red hairs	Black or very dark
FRUIT			
WINGS	Wing terminal	Seed body grading gradually into wing	Flat, completely surrounds seed body
SEED BODY	Cigar-shaped	Funnel-shaped	Slightly twisted, less than half the length of the fruit
TWIGS			
TEXTURE	Smooth and shiny, often with slight bloom, very brittle	Somewhat covered with downy hairs	Smooth, not shiny
COLOR	Gray or greenish-brown, inner bark bright brick red	Greenish-gray, inner bark cinnamon-colored	Pale gray, inner bark dirty white

*Specimens of green ash which lack hairs on the twigs or leaflets, but otherwise fit the above description, were formally designated as var. lanceolata. They are now designated under the species due to the many gradations of the hairiness character.

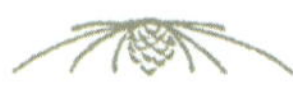

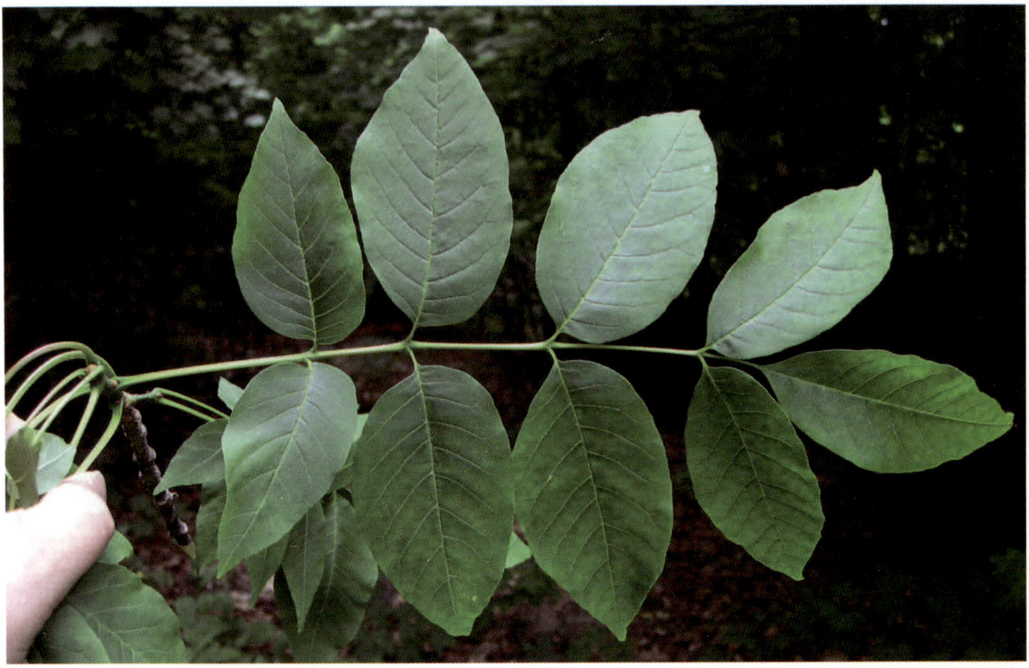

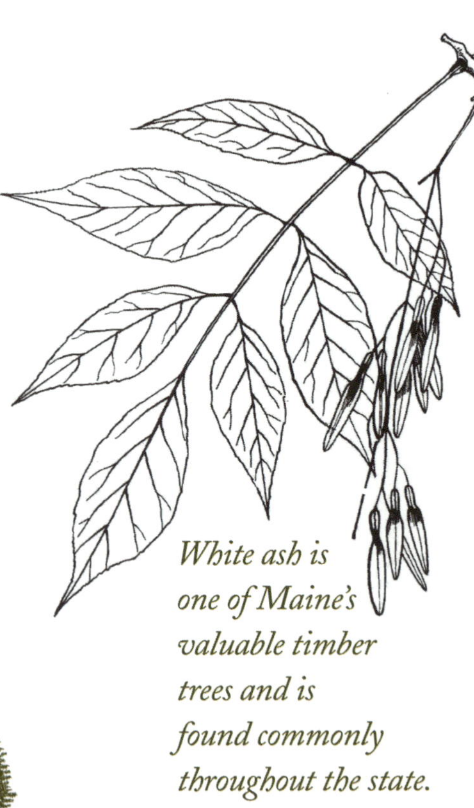

WHITE ASH *Fraxinus americana* L.

White ash is one of Maine's valuable timber trees and is found commonly throughout the state.

White ash is one of Maine's valuable timber trees and is found commonly throughout the state. Best growth occurs on rich, rather moist soil of low hills. It grows to a height of 60–70 feet and a diameter of 15–30 inches. The branches are upright or spreading, forming a narrow top in the forest.

The **bark** pattern resembles a woven basket; it is broken into broad, parallel ridges by deep furrows, and is dark brown or deep gray.

The **leaves** are opposite, 8–12 inches long and consist of 5–9 (usually 7) leaflets. The leaflets are 3–5 inches long, oval to lance-shape, borne on short stalks, edges remotely toothed towards the tip, dark green and often shiny on the upper surface. In fall, they turn to a soft, velvety purple.

The **fruit** is a single samara occurring in clusters. The seed body is cigar-shaped and has a terminal wing.

MAINE REGISTER OF BIG TREES 2008
White Ash
Circumference: 244"
Height: 95'
Crown Spread: 70'
Location: South Waterford

White ash twigs are hairless and have deeply notched leaf scars.

The **twigs** have a smooth, shiny bark which is grayish, greenish or maroon on the surface. The inner layer of the bark is brick red. The terminal buds are rounded or dome-shaped.

The **wood** is hard, strong and tough. It is used for agricultural implements, tool handles, oars, furniture, interior finish, dowels, pulp and firewood, and sporting goods including baseball bats, hockey sticks and snowshoe frames.

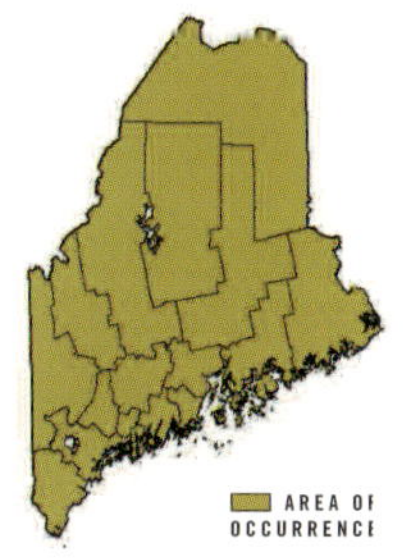

GREEN ASH *Fraxinus pennsylvanica* Marsh.

Green or red ash occurs over much of the state, particularly along the major rivers. It is not as abundant as the white and black ash, but is fairly common in central Maine. Sometimes mistaken for black ash, it grows near the banks of streams and lakes on rich, moist soil. It has stout branches that bend downward on older trees and form an irregular, compact head in the forest. It seldom exceeds a height of 50–60 feet and a diameter of 16–20 inches.

The quality of green ash wood is not as good as white ash.

Green ash twigs are often hairy and do not have deeply notched leaf scars.

MAINE REGISTER OF BIG TREES 2008
Green Ash
Circumference: 115"
Height: 63'
Crown Spread: 65'
Location: Mechanic Falls

The **bark** on the trunk of old trees is dark gray or brown, and firm and furrowed like that of the white ash.

The **leaves** are 10–12 inches long, opposite, with 7–9 leaflets borne per stalk. Leaflets are 4–6 inches long, entire or wavy, or sometimes toothed, particularly on the upper-half of the leaflets, yellow-green on the upper surface, hairy below and on the rachis, and oval to elliptical.

The **fruit** has a funnel-shaped seed body gradually blending into the terminal wing.

The current year **twigs** are greenish-gray and covered with numerous hairs, although sometimes there are no hairs. Inner bark is cinnamon red.

The **wood** is hard, heavy, fairly strong, coarse-grained and brittle. It is used in the same ways as white ash.

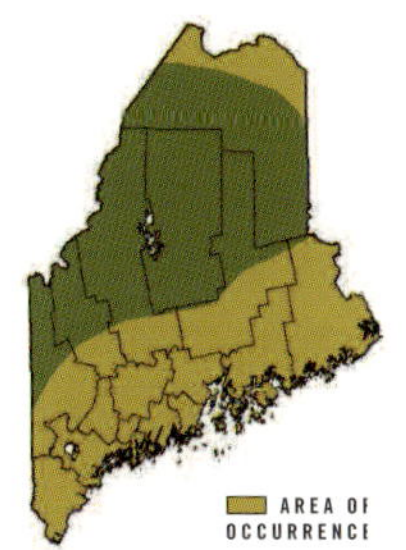

BLACK ASH *Fraxinus nigra* Marsh.

Black or brown ash occurs statewide. It grows almost entirely on rich, moist ground or in cold, wet swamps and along the banks of streams.

It is a tall, slender tree with a short, narrow head. It grows to a height of 50–60 feet and a diameter of 10–20 inches. The trunk is often without branches for a considerable distance from the ground.

The **bark** is gray to dark gray, corky and spongy, with more or less parallel ridges. It rubs off freely with the hand.

Black ash wood is used for interior finishing, cabinet work, baskets and, to a limited extent, pulp.

The wing of black ash fruit completely surrounds the seed body.

The **leaves** are 12–15 inches long, opposite, and have 7–11 leaflets that are 4–5 inches long, and without stalks except the one at the tip. Leaflets are lance-shape and have remotely-toothed margins. The upper surface is dark green. There are buff-colored hairs at the junction of the leaflets and rachis.

The **fruit** is a single samara occurring in clusters. The seed is flattened and completely surrounded by the wing.

The **twigs** are smooth, gray to olive-green. The buds are black or brown and pointed at the tip. The inner layer of the bark is dirty white.

The **wood** is coarse-grained, heavy, tough, durable and pliable. It is used for interior finishing, cabinet work, baskets and, to a limited extent, pulp. It the past it was used to make barrel hoops.

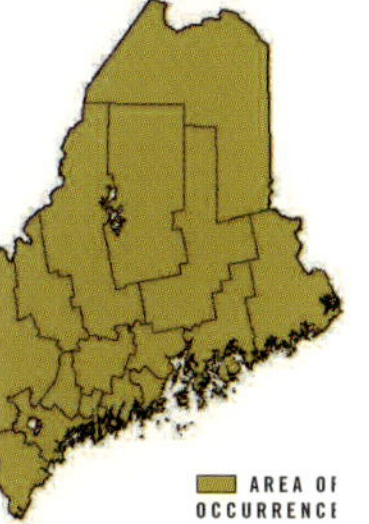

AMERICAN BASSWOOD *Tilia americana* L.

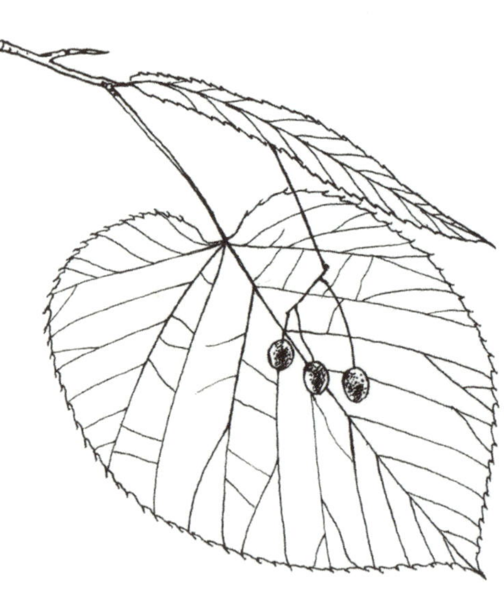

American basswood or linden occurs as scattered specimens throughout the state. It grows to a height of 50–70 feet and a diameter of 2–3 feet. The branches are slender, somewhat pendulous, comparatively small and numerous, forming a broad and rounded head.

The **bark** on the trunk of old trees is deeply and irregularly furrowed. On young trees, it is smooth or slightly fissured and has a grayish appearance.

The **leaves** are alternate, 5–6 inches long with uneven bases. They are broadly egg-shaped to heart-shaped in outline, and toothed; the upper surface

Light, soft, easily worked and carved, American basswood is used for molding, yardsticks, veneer, dowels, furniture, carvings and pulp.

The fruit of American basswood is attached to a distinctive leaf-like bract.

is dark green, while the lower is yellow-green and shiny.

The **flowers** are greenish-yellow, borne on a slender stalk that is attached to a rather long, yellowish, leaf-like bract. They are fragrant, contain an abundance of nectar and open in July.

The **fruit** is clustered, spherical, covered with short buff-colored hairs, woody and about as large as a pea. It remains attached to the leaf-like bract when it falls.

The **twigs** have a zigzag pattern and bright red buds.

The **wood** is light, soft, easily worked and carved. It is used for molding, yardsticks, veneer, dowels, furniture, pattern stock, carvings and pulp. Traditionally it was used to make butter box molds, dough bowls and other kitchen items that touched food.

In Germany, basswood is called the bee tree. Bees make an excellent grade of honey from the flowers. The young fruit and flowers ground into a paste make an excellent substitute for chocolate.

The **European linden** (*Tilia europaea* L.) and **Little-leaf linden** (*Tilia cordata* Mill.) are commonly planted as shade trees. They are smaller in height than our native species and with smaller leaves. Baxter Boulevard in Portland is lined with both of these species.

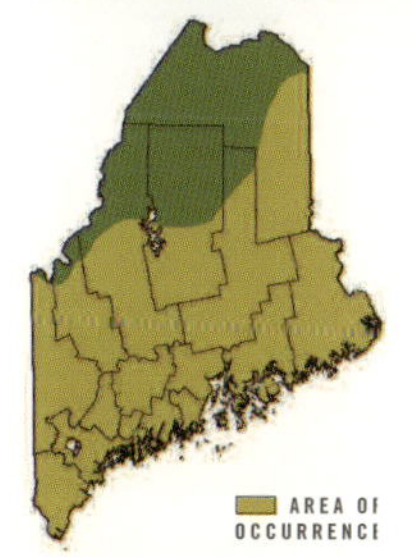

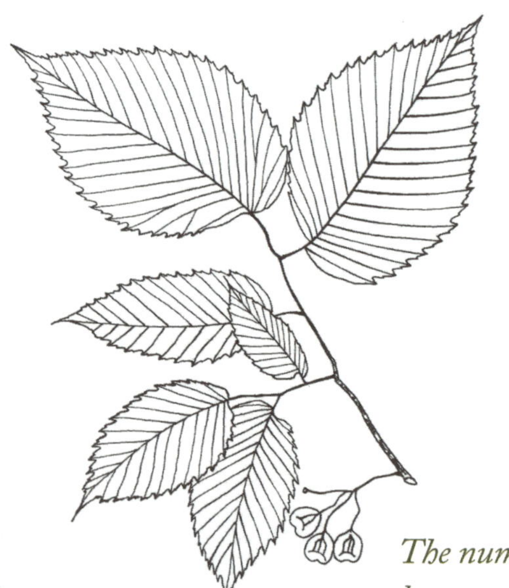

AMERICAN ELM *Ulmus americana* L.

American elm is one of our largest and most graceful trees; it occurs throughout the state, although its numbers have been severely reduced by Dutch elm disease. It is found most often on rich bottomland and moist soil along streams, but sometimes grows on higher ground. It grows quickly, attaining a height of 60–70 feet and a diameter of 2–4 feet.

The trunk often divides into numerous limbs, which form a vase-shaped or spreading, round-topped head with graceful, drooping branches.

The number of American elms in Maine has been severely reduced by Dutch elm disease.

The **bark** on the trunk is separated into broad ridges by deep fissures and is ashy-gray on the surface. It shows alternate layers of chocolate brown and buff coloration beneath.

The **leaves** are alternate, 3–6 inches long, with coarsely doubly-toothed margins and uneven bases. The upper surface is dark green and sandpaper-like.

The **flowers** appear in April before the leaves.

The **fruit** consists of a small, winged seed which ripens about the end of May, before the leaves have fully developed. It has a wide, open notch at the apex and a hairy margin.

The **wood** is spiral and coarse-grained, hard, heavy, strong, tough and hard to split. It is used for flooring, railroad ties and pulp. In the past it was used to make barrel hoops, barn stall flooring, door thresholds and wheel hubs.

Slippery elm, *Ulmus rubra* Muhl, has been recorded in Franklin and York counties, but these records are historical. A few specimens have been found in association with cultural settings, but it is not known if these populations are native or escaped. If it still occurs naturally in the state, it is undoubtedly quite rare. Slippery elm is most easily distinguished from American elm by the winter buds which are covered with rusty hairs. In the past, the inner bark of the slippery elm was chewed to relieve sore throats.

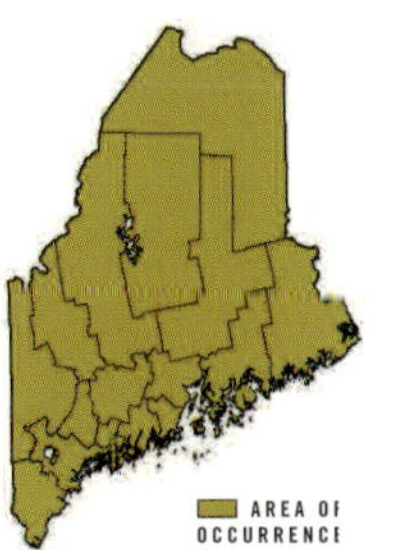

American elm twigs have a zigzag pattern and slightly flattened buds.

Loads of logs were "snubbed" when going downhill to prevent the horses from being overtaken by the load.

CHERRIES AND PLUMS *The Important Distinctions*

	Pin Cherry *Prunus pennsylvanica*	**Black Cherry** *Prunus serotina*	**Common Chokecherry** *Prunus virginiana*	**Canada Plum** *Prunus nigra*
BARK				
TEXTURE	Nearly smooth. Large horizontal lenticels show orange when rubbed.	Young trunks: prominent white lenticals. Older trunks: fissured and ridged.	Smooth with a pungent, disagreeable odor. Lenticels less prominent than on other Prunus species.	Lenticels yellowish
COLOR	Reddish-brown	Young trunks are black	Grayish-brown, with light-colored fissures	Dull reddish-brown to black
LEAVES				
GENERAL DESCRIP-TION	Long and tapering from base to tip. Widest in the lower ⅓; thin and firm textured with round teeth. Glands on stalk, and no hairs on midribs.	Elliptic/oblong, widest in the center, thick leathery and shiny. Underside of midrib near stalk end covered with rusty, brown hairs. Glands on stalk near blade. Margin has rounded teeth.	Obovate, widest in the terminal ⅓, sharply saw-toothed and without hairs, medium leathery in texture, glands on stalk and no brown hairs on midrib.	Ovate or obovate tapering abruptly into a long thin point. Teeth rounded. Glands on stalk.
TWIGS				
SHAPE	Very fine	Waxy	Medium slender	Thorns common on older twigs
COLOR	Red and reddish-brown	Red-brown with a lighter or greenish margin	Gray or purplish-brown	Current growth gray, older growth darkening to black
ODOR	Slight cherry odor	Sharp, pungent smell when broken	Strong, pungent bitter-almond odor	None
BUDS				
SHAPE	Football-shaped with a longitudinal furrow	Ovate, flattened	Cone shaped, slender – pointed, side buds not flattened	Cone shaped, pointed
COLOR	Red-brown	Red-brown with a lighter or greenish margin	Purple to tan pattern	Gray-brown
FRUIT				
COLOR	Bright Red	Black	Deep red to purple	Light red to yellow
SIZE	¼ inch diameter	½ inch diameter	¼–½ inch diameter	1 inch diameter
ARRANGE-MENT	Hang in umbellate or corymbose clusters	Produced in a raceme, the individual fruit have a persistent basal disc	Produced in racemes, basal disc not persistent	Football-shaped with a longitudinal furrow

The twigs and branches of cherry and plum trees may be distorted by the 'black knot' fungus (Dibotryon morbosum (Schw.)Th. & Syd.).

PIN CHERRY *Prunus pensylvanica* L. f.

Pin cherry is not used commercially and has little value except as a protection and cover for the soil on recent clearings or burned areas.

Pin or fire cherry is a small tree, seldom growing taller than 25–30 feet in height and 6–10 inches in diameter. It has slender, horizontal branches and a narrow, somewhat rounded head. It is common throughout the state, but has little value except as a protection and cover for the soil on recent clearings or burned areas.

The **bark** on the trunk of old trees is dark red-brown and broken into thin plates. Bark on young trees is smooth and reddish-brown. The inner bark is slightly aromatic and very bitter. The large lenticels show orange when rubbed.

The **leaves** are alternate, narrow to oblong, widest in the lower third, rather sharp pointed, finely and sharply toothed on the edges, bright green and shiny above, without hairs below, 3–4 inches long, bitter and aromatic, with glands on petiole.

The **flowers** are white, in clusters of four or five, and appear in May when the leaves are only half grown. The **fruit** is bright red, almost translu-

cent, pea-sized, globular and ripens from the first of July to August. In the past, it was used to make jams and jellies.

The **twigs** are shiny, reddish-brown and very slender. The buds are small, ovoid, reddish-brown and clustered at the end of twigs. They are commonly distorted by a black, warty, fungus growth called "black knot."

The **wood** is coarse-grained, soft and light. It is not used commercially. The gum is edible.

Pin cherry produces abundant 5-petaled flowers every spring.

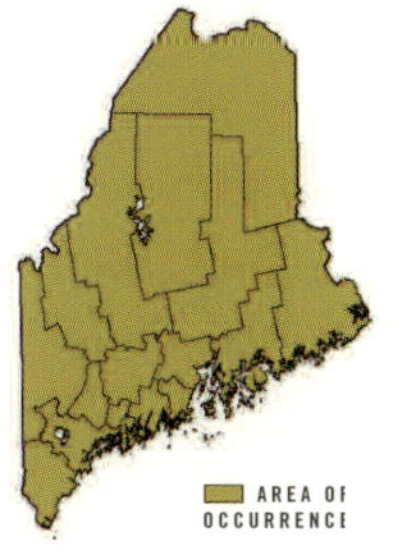

BLACK CHERRY *Prunus serotina* Ehrh.

Black cherry is widely distributed throughout the state. In some parts of the country it is an extremely valuable timber tree, but in Maine does not often grow to sufficient size. It grows on a variety of soils, but makes rapid and best growth on rich, moist land. It has a narrow head, small horizontal branches, and attains a height of 40–50 feet and a diameter of 10–20 inches.

Black cherry is one of our most valuable timber trees, although not abundant in sufficient size.

The **bark** on the trunk of young trees is red-brown to black and rather shiny with prominent white lenticels. On older trees, the bark is broken into small irregular plates.

The **leaves** are alternate, elliptic, oblong, widest at the center, finely toothed, dark green, shiny, thick, somewhat leathery and 2–5 inches long. The underside of the midrib near the stalk end is covered with rusty brown hairs.

The **flowers** are produced in many flowered racemes 4–5 inches long that appear at the end of May or in early June when the leaves are half-grown. The **fruit** is in drooping racemes, dark purple or almost black when ripe, ¼–½ inch in diameter, and globular in shape. It ripens from June to October and is an important wildlife food.

The **twigs** give off a pungent odor when broken, and the bark has a bitter taste. As with pin cherry, twigs and branches are commonly distorted by a black, warty, fungus growth called "black knot."

The **wood** is rather hard, close-grained, light, strong and easily polished. It is used for furniture and cabinetmaking, interior finishing, woodenware, veneer and plywood. It is valued as firewood due to its high heat value and fragrance.

The rusty hairs along the midvein on the back of the leaf distinguish black cherry.

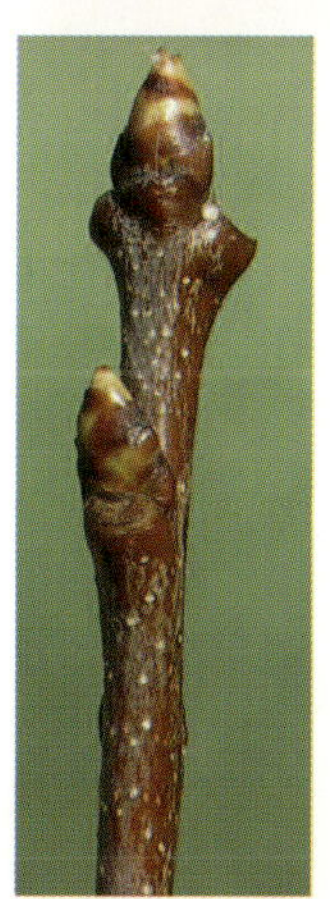

Black cherry twigs have a bitter taste and give off a pungent odor when broken.

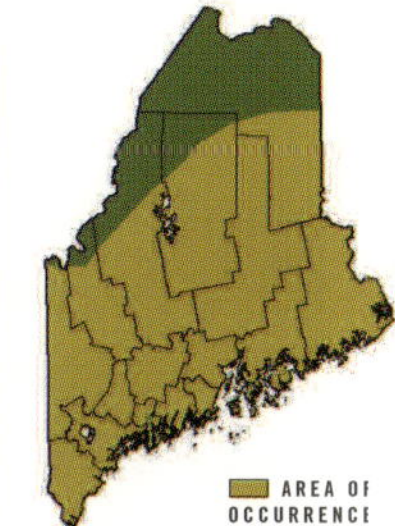

MAINE REGISTER OF BIG TREES 2008
Black Cherry
Circumference: 148"
Height: 52'
Crown Spread: 60'
Location: Falmouth

COMMON CHOKECHERRY
Prunus virginiana L.

Common chokecherry wood is heavy and hard, but not strong, and is not used commercially.

Common chokecherry is a shrub or small tree that occurs throughout the state, especially along fencerows in farming communities. It occasionally is 25 feet high and 6 inches in diameter.

The smooth, grayish-brown to black **bark** is usually marked by long, light-colored fissures and has a disagreeable scent.

The **leaves** are alternate, dull, widest at the terminal one-third, 2–4 inches long, finely-toothed, medium-leathery in texture and at maturity are without hairs.

The **flowers** appear from the first of May to June on slender stalks in racemes.

Chokecherry twigs have cone-shaped buds and give off a strong odor when broken.

The **fruit** ripens from July to September, and is about ¼–⅓ inch in diameter, at first bright red, turning at maturity to dark red or nearly black. It is slightly astringent, but edible.

The **winter buds** are strictly cone-shaped, slender and pointed with a definite purple and tan pattern on the scales. Side buds are not flattened as in black cherry.

The **twigs** have a strong, pungent, skunk-like odor when broken and, as with pin cherry and black cherry, are frequently distorted by a black, warty, fungus growth called "black knot."

The **wood** is heavy, hard but not strong, and is not used commercially.

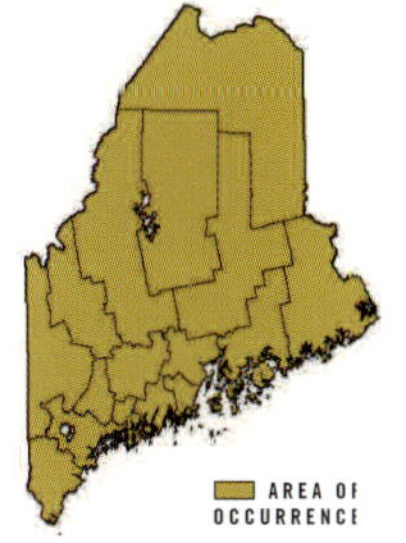

The Canada plum's fruit, which ripens in the latter part of August, is edible.

CANADA PLUM *Prunus nigra* Ait.

Canada or red plum, has been planted as an ornamental and is found occasionally throughout much of the state. It does not occur in densely forested areas; rather it usually occurs in thickets along field edges. It is seldom over 8 inches in diameter and 30 feet high. The twigs and branches of cherry and plum trees are distorted by the black knot fungus, *Apiosporina morbosa.*

The **bark** is thin, dull reddish-brown to black; it peels in thin papery scales, exposing the shiny reddish-brown, inner bark.

The **leaves** are alternate, obovate, and taper at the apex to a long, sharp point. Leaves are dark green on the upper surface, lighter below; the margin has glandular, rounded teeth.

The **flowers** are white, and appear early in spring before the leaves in groups of three or four on slender stalks. The edible **fruit** ripens the latter part of August, is football-shaped and furrowed along one side, and has an orange-red skin and yellow flesh. The single stone is flattened and slightly grooved on the edges.

The **twigs** and **branches** often have thorns. The buds are brown to gray and are without hairs. The **wood** is not used commercially.

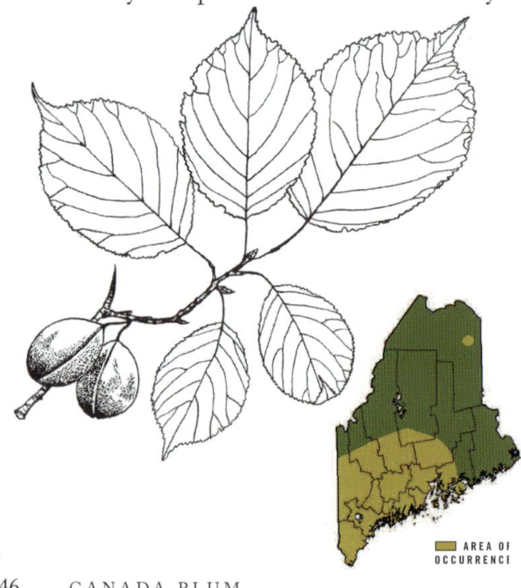

HAWTHORN *Crataegus spp.* L.

Hawthorn, or thorn-apple, occurs in Maine as a low spreading tree or shrub that rarely reaches a height of more than 15–18 feet. There are approximately 22 different species found in the state. Hawthorns can usually be recognized by the small apple-like fruits and the thorns on the branches. In the past, hawthorns were planted as hedges in place of fencing.

The **bark** is dark brown to ashy gray and somewhat scaly.

The **leaves** are alternate, doubly-toothed, and usually somewhat lobed, thin and dark green.

The **flowers** appear about the first of June in flat, showy white clusters.

The **fruit,** which is ¾ inch in diameter, resembles a small apple. The flesh is thin, mealy and encloses 1–5 rounded nutlets. It is used for jellies and bird food.

The **twigs** are slender, rigid and usually armed with long thorns. They form a compact crown due to their zigzag method of growth.

The fruit of the hawthorn is used for jellies and bird food.

The **wood** is heavy, hard and close-grained. It is used to some extent for handles and other small articles.

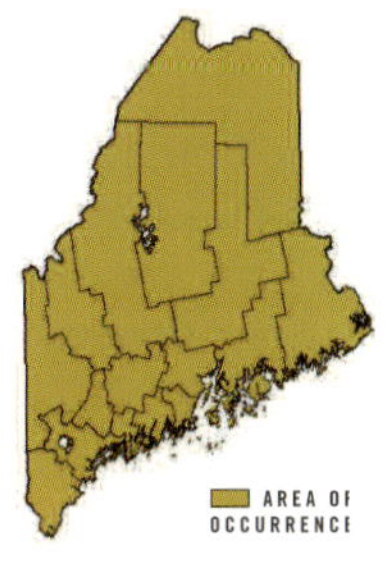

SERVICEBERRY *Amelanchier spp.* Medik.

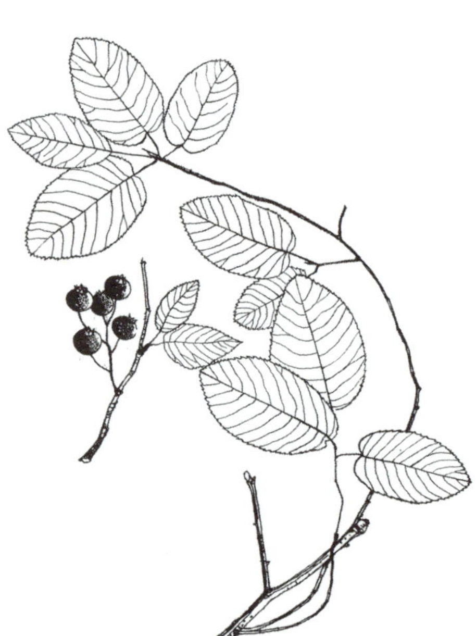

Serviceberry wood is occasionally used for tool handles, small implements and fishing rods.

Approximately seven species of serviceberry or shad bush grow as shrubs or small trees in Maine. Of these, two species—Allegheny serviceberry *Amelanchier laevis* Wieg. and downy serviceberry *Amelanchier arborea* (Michx. f.) Fern.—commonly grow to be small trees 30–40 feet in height and 6 to 8 inches in diameter. Allegheny serviceberry, is the more common of the two. They are both found in open hardwood stands or along the margins of open areas throughout much of the state.

The **bark** of serviceberry is smooth, gray to light violet-brown with darker vertical stripes; older bark is slightly fissured longitudinally and twisted.

The **leaves** of Allegheny serviceberry are half grown at flowering time, and have a reddish or purplish tinge. When downy serviceberry leaves are just unfolding, they are green and densely hairy beneath. Mature leaves of both species are alternate, dark green

Serviceberry buds are long and sharp-pointed; the lateral buds hug the twig.

above and lighter green below, 1½–3 inches long, 1–1½ inches wide, elliptic to ovate with a rounded or heart-shaped base.

The **flowers** are white and sweet-smelling with 5 petals. The serviceberry flowers before other trees and is very easy to spot along the edges of fields and streams in spring.

Serviceberry **fruit** is berry-like, ripens in early summer, is ⅓–½ inch in diameter, and red to dark purple when mature and edible. Serviceberry **twigs** are slender, red-brown and finely hairy when young, becoming smooth as the twigs grow. The buds are long, sharp pointed, reddish or pinkish, and filled with silky hairs.

The **wood** is occasionally used for tool handles, small implements and fishing rods. It is heavy, hard, strong, close-grained and dark brown tinged with red.

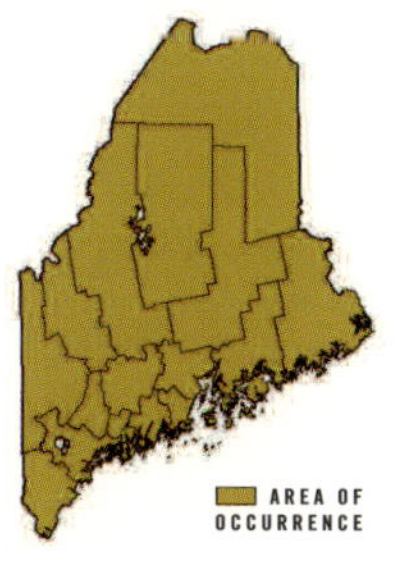

Mountain Ash *Sorbus spp.*

Showy mountain ash is usually better balanced in outline than the American mountain ash and has a well-rounded crown.

There are two native species of mountain ash found in Maine: the American mountain ash *Sorbus Americana* Marsh, also called round-wood, and the showy or Northern mountain ash *Sorbus decora* (Sarg.) Schneid.

American mountain ash occurs statewide; it is not a true ash, but is closely related to the apple. It rarely reaches over 20 feet in height. It is particularly common in mountainous regions and along the coast. The leaves are alternate, compound 13–17 inches long, tapered, and have 11–17 finely toothed leaflets. The leaflets are 2–4 inches long, ⅝–1 inch wide, and without hairs.

The small creamy-white **flowers** are borne in cymes. The berry-like **fruit** is bright red, and about ¼ inch in diameter. These remain on the tree late into the winter; they make good bird

food. In the past, they were sometimes used as an astringent in medicine. The bud scales are hairless and sticky. The pale brown **wood** has little value because it is soft and weak.

Showy mountain ash is most commonly found in northern and western parts of the state. It is usually better balanced in outline than the American mountain ash and has a well-rounded crown. The **leaves** are alternate, compound, and differ from the preceding species in having leaflets which are only 1½–3 inches long, and ⅝–1⅝ inches wide. The **fruit** is larger, up to ½ inch in diameter, and matures later in the season. The outer **bud** scales are sticky; the inner scales are hairy.

This photo is of American mountain ash fruit. Showy mountain ash fruit is larger.

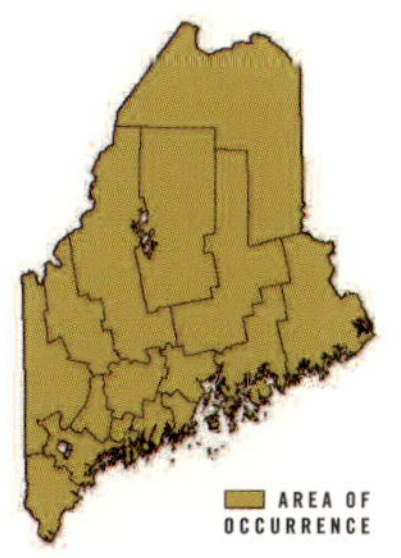

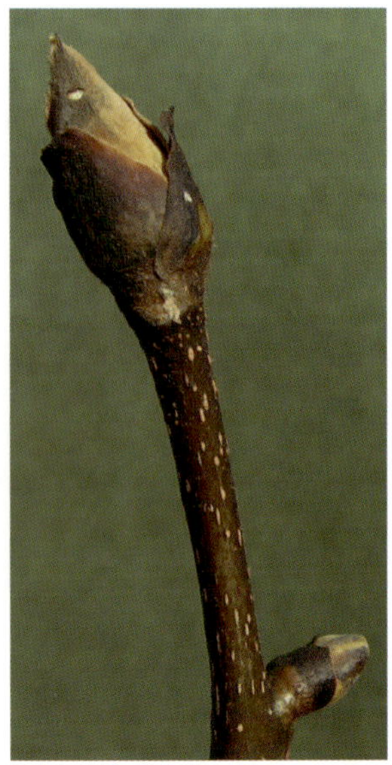

Shagbark Hickory
Carya ovata (P. Mill.) K. Koch

Shagbark hickory is most commonly found in southern Maine on moist but well-drained soil. It has a cylindrical head and a straight, gradually tapering trunk. It reaches a height of 70 feet and a diameter of 2 feet.

The **bark** is light gray on the trunk and separates into long, loose plates, giving it a shaggy appearance.

The **leaves** are compound, alternate, 8–14 inches long; most often there are 5 leaflets, rarely seven. The 3 terminal leaflets are the largest. Leaflet margins are serrate.

The **fruit** has the thick outer husk deeply grooved at the seams. The husk separates along these grooves when ripe. The fruit is globose and is borne singly or in pairs. The edible kernel is sweet. The **twigs** are hairy or smooth and olive-gray to dark red-brown. Pith is star-shaped in cross section. Bud scales are hairy.

Shagbark hickory wood is primarily used to make pallets.

The **wood** is very strong, close-grained, heavy, hard, tough and flexible. It was formerly used in the manufacture of agricultural implements, axe and tool handles, carriages and wagons, especially the spokes and rims of the wheels. Its principal uses are now pallets, pulp and firewood.

Shagbark hickory is easily distinguished by its bark, which separates into long, loose plates.

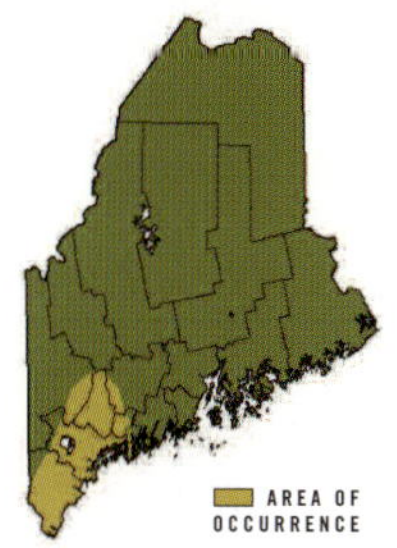

BITTERNUT HICKORY
Carya cordiformis (Wangenh.) K. Koch

Although common further south, bitternut hickory is rare in Maine, occurring only in the extreme southwestern corner of the state at the southern tip of York County. Bitternut hickory will grow on a variety of sites, but makes its best growth on moist bottomland soils.

The **bark** of young trees is silvery-gray and smooth; older trees have gray bark with tight, shallow, interlacing furrows. The bark remains tightly attached on old trees and does not become shaggy. The **leaves** are 8–10 inches long, alternate, pinnately compound with 7–9 leaflets. The terminal leaflet is similar in size to the adjacent ones.

The **flowers** occur in spring; male flowers are in catkins and female flowers are in a terminal spike. The **fruit** is a nut; it is nearly round and only slightly flattened. It is covered by a thin green husk with 4 small wings descending from a sharp point to the middle. As the name bitternut implies, the meat is very bitter and not eaten by humans, although some wildlife utilize it as food. The **twigs** are somewhat stout (although much less so than other hickories) and have distinctive sulfur-yellow buds.

The **wood** is hard; further south, it used for making tool handles, furniture, paneling and pallets, as well as for fuel. It is a choice wood for smoking meats. In Maine, due to its rarity, the wood is not used commercially.

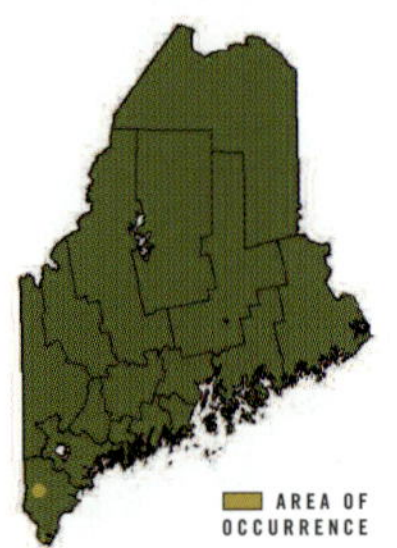

BLACK WALNUT *Juglans nigra* L.

Black walnut is not native to Maine, but is planted occasionally as an ornamental tree. In forested situations in its native range, it can grow to be up to 100 feet tall with a long straight trunk free of branches. In Maine, it is usually planted in the open and exhibits an open-grown form with wide-spreading branches. Black walnut's natural range extends over a large portion of the eastern United States from western Vermont and Massachusetts to southeastern South Dakota, south into Texas and the Florida panhandle.

The **bark** is brown, with furrowed ridges forming a diamond pattern. If the bark is cut with a knife, the cut surface will be dark brown. The leaves are alternate, pinnately compound 12–24 inches long with 10 24 leaflets; a terminal leaflet is often lacking. The **fruit** is round and composed of a nut enclosed in a thick green husk. The **twigs** are stout, light brown, with a chambered pith. The **buds** are large and tan.

The **wood** is so valuable that, in some parts of the country, trees have been stolen in the dead of night from front lawns and city parks. It is a rich, dark brown and takes a good polish, making it valuable for furniture, cabinets and gunstocks. Much of the wood harvested today is turned for veneer.

The **nuts** are edible, but must be gathered before the animals harvest them all. Ground nut shells have had numerous uses, including as a carrying agent for insecticidal dusts and for cleaning aircraft engine parts; while the fruit husks have been used to make fabric dye.

BUTTERNUT *Juglans cinerea* L.

Butternut, also known as white walnut, occurs naturally or in cultivation to some extent statewide. It grows on rich, moist soil and on rocky hills, especially along fencerows. It frequently has stout, spreading limbs extending horizontally from the trunk to form a low, broad, rounded head. It grows to 30–40 feet high and a diameter of 1–2 feet. Currently, butternut is under severe threat from butternut canker, *Sirococcus clavigignenti-juglandacearum*. This fungus was most likely introduced from outside of North America and is now killing butternuts throughout much of Maine.

The **bark** of young trees and of the branches is gray. On old trees, it is broadly ridged on the trunk and light brown.

The **leaves** are compound with a terminal leaflet, alternate, 15–30 inches long, and consist of 11–17 leaflets. The leaflets have serrate margins.

The **fruit** is composed of a nut enclosed by a fleshy husk covered with sticky hairs. It is about 2½ inches long and oval shaped. Fruit is produced in drooping clusters of 3–5. The nut is thick-shelled with sharp ridges on the surface. American Indians used the oil from the nuts to make butter. Brown dye was made from the husk.

The **twigs** are stout, greenish and hairy, with chocolate-brown, chambered pith. The large leaf scars have a conspicuous, buff-colored hairy pad at the top; the buds are also hairy.

The **wood** is coarse-grained, light, soft and weak. It is sometimes used for furniture and cabinetwork and takes a high polish.

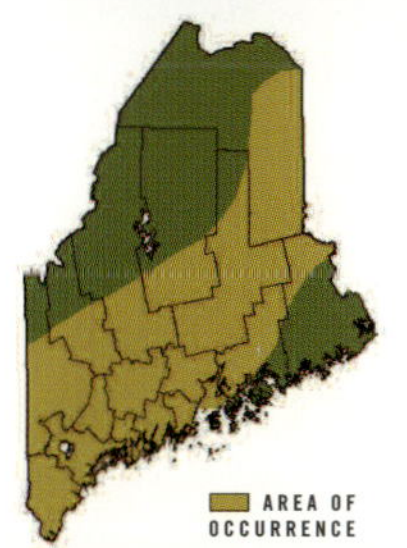

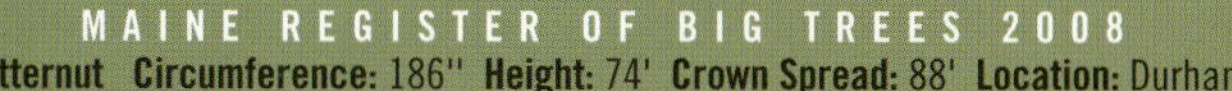

HORSECHESTNUT
Aesculus hippocastanum L.

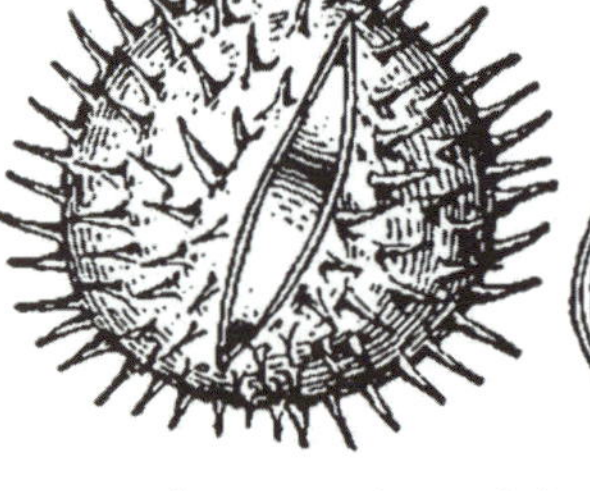

Not related to the native chestnut, the horsechestnut comes from Asia and the Balkan Peninsula and is generally planted as a shade and ornamental tree.

It is symmetrically round or oval in outline with a stiff branch habit. The tips of the branches curve slightly when mature. It has heavy, luxuriant, deep green foliage which changes to bronze in early autumn. The large, opposite **leaves** with 5–7 leaflets, are arranged palmately on a single stalk; and distinguish it from any of Maine's native trees. With the pyramids of white **flowers** blossoming in the early spring and the large, bur-like, leathery husk enclosing one or more smooth, mahogany-colored **nuts**, the horsechestnut is not easily confused with any other species. The nuts are poisonous when ingested. It makes a good shade tree, but requires rich soil for best development. It is prone to a leaf blight.

The **buds** are large, sticky and nearly black. The **wood** is soft, light and close-grained. In Europe, it is used for carving and veneer. In the past in the U.S., it was burned as firewood.

AMERICAN SYCAMORE
Platanus occidentalis L.

There are historic records of American Sycamore occurring along streams and on rich bottom lands in southern Maine. Currently there are no known native populations in the state. However, sycamore is planted here as an ornamental. Farther south and west it grows to be an enormous tree, often 4-6 feet in diameter and 120 feet tall, trees in Maine however do not attain great size.

The **bark** on the trunk and large limbs is greenish-gray and flakes off in broad scales exposing white patches beneath.

The **leaves** are simple, alternate, 3–5 lobed and light green. The base of the leaf-stalk is hollow, swollen and covers the bud.

The **fruit** head generally occurs singly, is round and about 1 inch in diameter. It contains a large number of small wedge or shoenail-shaped nutlets, and usually remains on the tree until spring.

The **twigs** are zigzag in shape and are encircled by conspicuous stipules. The winter buds have a single, wrinkled, cap-like scale.

The **wood** is hard, firm, very perishable when exposed to the weather, and liable to warp. In the past, it was used for furniture and interior finish of houses.

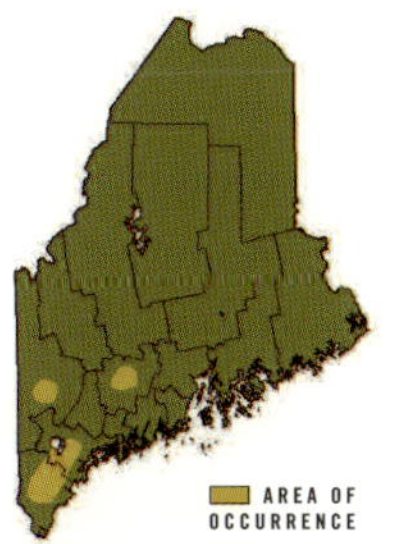

Black Tupelo *Nyssa sylvatica* Marsh.

While black tupelo wood is heavy, fine-grained and very tough, it is not durable and is used principally for pulp.

Black tupelo, or blackgum, is found in Sagadahoc, Androscoggin, Cumberland and York counties and as far north as Southern Oxford County and Waterville in Kennebec County. However, it is not commonly found except in very wet areas. Trees 2 feet in diameter are found in the town of Casco on an island in Sebago Lake. Large specimens have also been reported on the south side of Pleasant Mountain in Denmark on a flat, open, wet area. Easily distinguished at a distance by its numerous slender horizontal branches, the tree rarely reaches more than 50 feet in height. It occurs in rich moist soils, such as swamps or borders of rivers. Black tupelo can live to a very old age. Trees over 500 years old have been found in New Hampshire.

The **bark** on young trees is smooth, grayish and flaky, later becoming reddish to grayish-brown. On old trees, it forms coarse blocks or ridges.

The **leaves** are alternate, oval to obovate, 2–5 inches long, wedge-shaped at the base and pointed at the tip. The edges are usually entire. The leaves are dark green, shiny above, occasionally hairy below, and turn bright crimson in autumn.

The **fruit** is dark blue, fleshy, approximately ½ inch in length, and borne in clusters of 1–3 on long, slender stems. The fruit has an acid taste, but is edible.

The **twigs** are moderately stout with a diaphragmed pith.

The **wood** is heavy, fine-grained, very tough but not durable. It was formerly used for the hubs of wheels and soles of shoes. It is now used principally for pulp.

Black tupelo has a characteristic horizontal branching pattern.

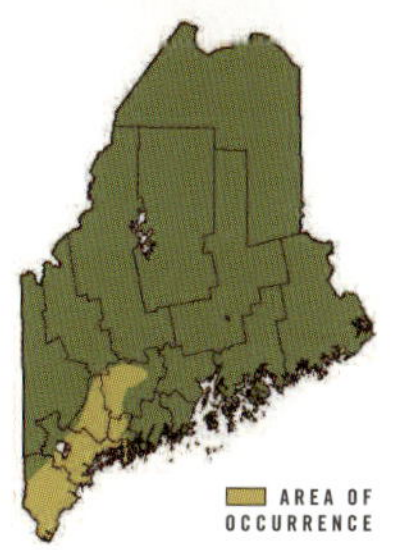

BLACK LOCUST *Robinia pseudoacacia* L.

The locust is a rapid grower, frequently attaining a height of 20 feet in 10 years, but increasing much more slowly thereafter.

Black locust is not a native of this state, but is extensively planted. It is abundant in some localities, and is found mostly near dwellings or on abandoned farmlands, where it often becomes naturalized. The locust is a rapid grower, frequently attaining a height of 20 feet in 10 years, but increasing much more slowly thereafter. It reaches a height of about 50 feet and a diameter of 8–20 inches. The branches are small, brittle, occasionally multi-angled, and at first are armed with stipular spines. The top is narrow and oblong. It is one of the last trees to send out foliage in the spring.

The **bark** on old trees is dark brown, deeply furrowed and broken into small scales.

The **leaves** are alternate, once compound, 8–14 inches long, with 7–19 leaflets that are about 2 inches long with an entire margin and a slightly notched tip.

The **flowers** are borne in loose racemes 4–5 inches long. Showy and very fragrant, they appear in June.

The **fruit** is a smooth, flat, dark purplish-brown pod about 3–4 inches long, containing 1–8 bean-like seeds.

The **wood** is heavy close-grained, strong, and very durable when in contact with the soil. It is used for fence posts, firewood and planking for boats. In the past, it was used to make pegs for use with glass insulators.

The buds of the black locust are almost completely hidden.

MAINE REGISTER OF BIG TREES 2008
Black Locust **Circumference:** 205'' **Height:** 80' **Crown Spread:** 66' **Location:** Belfast

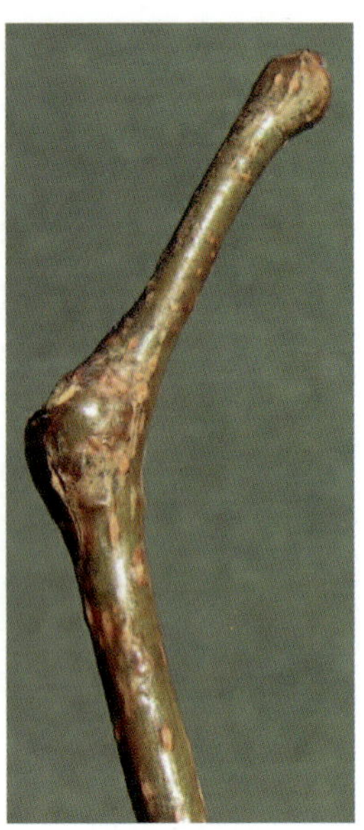

HONEYLOCUST *Gleditsia triacanthos* L.

Honeylocust is not native to Maine, but has been frequently planted in urban areas in the southern and central portions of the state. The trees most commonly planted and those that have escaped cultivation in Maine are a thornless variety, *Gleditsia triacanthos* f. *inermis* (L.) Zabel.

Honeylocust has somewhat pendulous, slender, spreading branches that form an open, broad, flat-topped head. It attains a height of 75 feet and a diameter of 20 inches. Simple or (usually) three-forked spines, 1½–3 inches long or longer, occur on the branches and trunk; but spines are lacking on the commonly planted variety.

The **bark** is divided into long, narrow ridges by deep fissures; and the surface is broken into small scales that are persistent. The **leaves** are alternate, both once and twice compound, 4–8 inches long and have from 18–28 leaflets. The margins of the leaflets are finely blunt-toothed.

The **flowers** are borne in slender clusters 2–2½ inches long. They appear in June when the leaves are about fully-grown. Staminate and pistillate flowers are produced separately on the same tree.

The **fruit** is a shiny, reddish-brown, flattened pod 8 inches or more in length. The pod is curved, with irregular wavy edges, and is often twisted. The walls are thin and tough.

The **twigs** are smooth and distinctly zigzag in shape. Winter buds barely protrude from the leaf scar.

The **wood** is coarse-grained, hard, strong, and very durable in contact with the soil. It is used for firewood and boat decking in Maine. In the past, it was used to manufacture the wooden pegs that glass insulators were screwed onto when glass insulators were used with telegraph, telephone and electrical power lines.

SASSAFRAS *Sassafras albidum* (Nutt.) Nees

Sassafras occurs in southern Maine in eastern Cumberland, southern Oxford and York counties, and is sometimes planted for ornament. Excellent specimens may be seen in the York Village cemetery.

The **bark** on young stems is thin and reddish-brown. On older stems, it becomes thick and scaly. The inner bark is very fragrant and sometimes chewed.

The **leaves** are alternate, very hairy when they first appear, losing the hair at maturity except on the midrib. They are light green and of 3 shapes: entire, mitten-shaped and three-lobed.

The **flowers** open in early spring with the first leaves, in racemes containing about 10–15 flowers.

The **fruit** ripens in September and October and is a blue, lustrous drupe that is supported on a fleshy, red stalk.

The **twigs** are green in color, smooth and aromatic when broken.

The **wood** is soft, weak, brittle, very aromatic, light brown and very durable in the soil. Historically, the roots and bark were distilled for oil of sassafras, used to perfume toiletries. The oil has been banned from use in foods in the US.

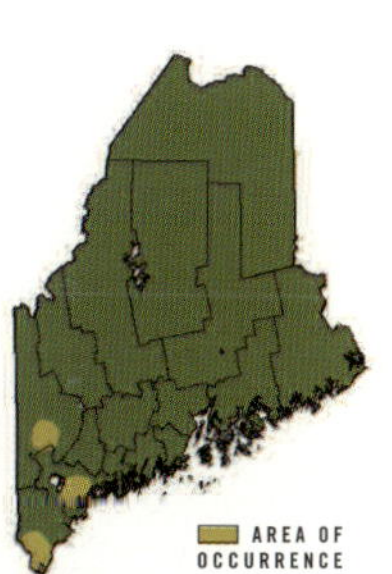

MAINE REGISTER OF BIG TREES 2008
Sassafras **Circumference:** 82" **Height:** 66' **Crown Spread:** 21' **Location:** York

NANNYBERRY *Viburnum lentago* L.

Nannyberry occurs statewide as a shrub or small tree reaching a height of 10–30 feet. It frequently is found growing in moist soils, often along the borders of swamps or streams.

The **leaves** are opposite, ovate, abruptly pointed, with fine sharp teeth. The upper surface is a lustrous deep green. The undersurface is lighter. The petiole is conspicuously flanged with a warty, wavy margin.

The dark blue **fruit** ripens in fall. It is about ½ inch long, ellipsoid, edible, sweet, tough-skinned, with a nipple-like tip. The fruit occurs in small drooping clusters on red-stemmed stalks, and does not shrivel or shrink when ripe.

The terminal **buds** are shaped like a pair of rabbit ears and bulge at the base. The 2 large bud scales extend beyond the end of the bud. They are nearly smooth and are purplish-brown to lead-colored. The smooth **twigs** of the season are gray to gray-brown. The **wood** is orange-brown and emits an unpleasant odor.

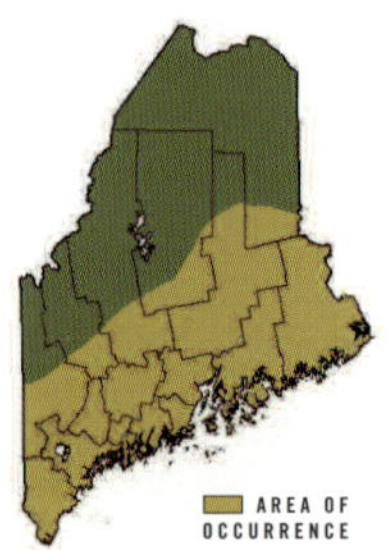

WITCH-HAZEL *Hamamelis virginiana* L.

Witch-hazel occurs as a small tree or shrub in most parts of Maine except in the far north. It is found on borders of the forest in low rich soil or on rocky banks of streams.

The **bark** is gray-brown and somewhat scaly on older stems. The **leaves** are alternate, broadly obovate, non-symmetrical at the base, and have a wavy margin.

It has bright yellow **flowers** with thread-like petals in autumn or early winter. The **fruit** is a woody capsule, usually two in a cluster. The seeds are discharged fiercely when ripe.

The **twigs** are gray, zigzag, with gray or rust-colored hair and scalpel-shaped buds.

An extract from the bark is mixed with alcohol and used as an astringent.

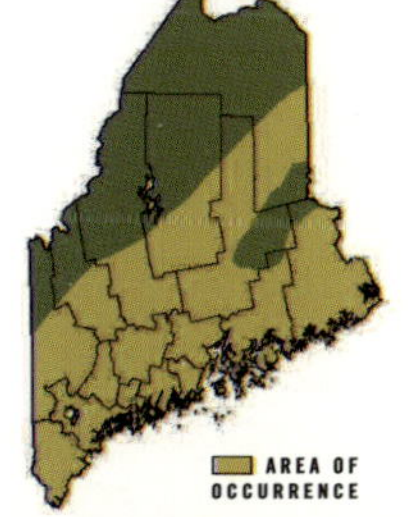

MAINE REGISTER OF BIG TREES 2008
Witch-hazel Circumference: 18'' **Height:** 32' **Crown Spread:** 17' **Location:** Rockport

STAGHORN SUMAC
Rhus hirta (L.) Sudworth

Staghorn sumac is a shrub or small tree that grows throughout most of the state. It can grow to about 25 feet tall and about 8 inches in diameter, although it is usually smaller. Occurring mostly on disturbed sites such as road sides and old fields, staghorn sumac sprouts readily from the roots. It often forms thickets that have a characteristic domed-shaped appearance with the tallest stems in the center. Unlike the unrelated poison sumac, staghorn sumac is not poisonous to the touch.

The **bark** is grayish-brown and has numerous lenticels. The **leaves** are 16–24 inches long, alternate, pinnately compound with 11–31 opposite, serrate leaflets. The leaves turn a brilliant red in fall. The **flowers** form in early summer in large, compact, yellow panicles 2–8 inches long. The **fruit** ripens in August as a spire of showy, red, velvety berries that often remain into the winter. The **twigs** are stout and very hairy. The winter **buds** are not covered by scales.

The **wood** has a greenish cast with dark grain. It is not used commercially except for small specialty items. Because the wood has a chambered pith which can be easily cleaned out, it was used for sap spiles (tubes for collecting sap in a sugarbush).

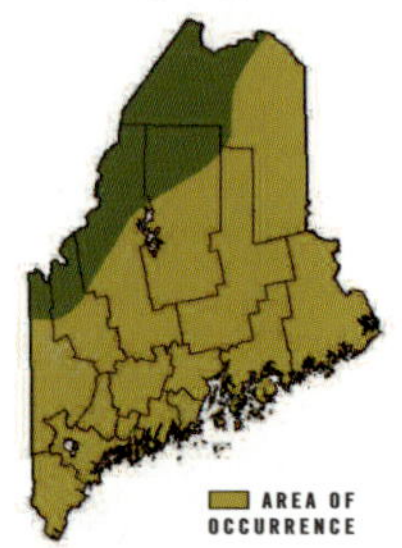

Mountain Laurel *Kalmia latifolia* L.

Mountain laurel is an erect-stemmed low shrub or small tree that grows in rocky woods or on low ground. Mountain laurel occurs rarely in southern and western Maine and is listed as a species of special concern.

The **leaves** are evergreen, green on both sides, elliptical, up to 3 inches long and 1 inch wide. They are flat, thick and leathery with an entire margin, and narrowed at both ends. Arrangement is mostly alternate, grouped at the tip of the twig, sometimes opposite and rarely in threes.

The **flowers** are pink with variations possible. They are borne in erect, terminal clusters.

The **fruit** are globose, woody capsules borne on erect, hairy, sticky stalks that are many times longer than the diameter of the capsules. The capsules have long, persistent styles.

The **twigs** are rounded and sticky at first, but later become smooth.

ROSEBAY RHODODENDRON
Rhododendron maximum L.

Rosebay rhododendron, or great laurel, is a shrub or straggling tree up to 30 feet high. It is a very rare species found locally in parts of Somerset, Franklin, Cumberland and York counties in damp woods or near pond margins. It is listed as a threatened species in Maine.

The **leaves** are evergreen, ovate to oblong, alternate, entire, 4–8 inches long, thick and leathery, with the margin frequently rolled under. They are smooth and dark green above, pale below. The **flowers** are bell-shaped and occur in dense clusters. They are generally white with a pinkish tinge with other variations possible. The **fruit** is an oblong, woody capsule covered with sticky hairs. It is borne terminally in erect clusters on stalks several times longer than the capsule. The **twigs** are hairy.

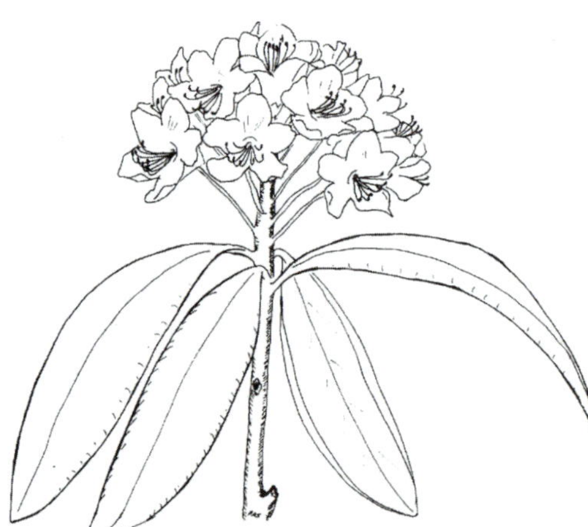

Rosebay rhododendron is listed as a threatened species in Maine.

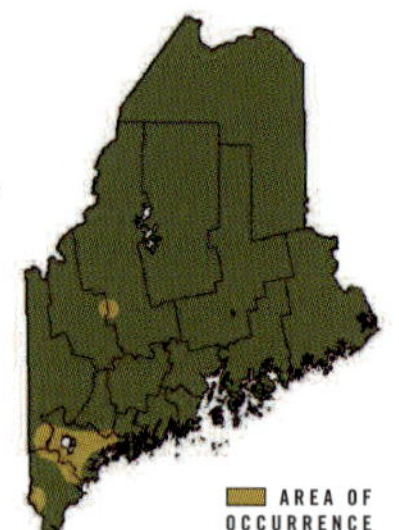

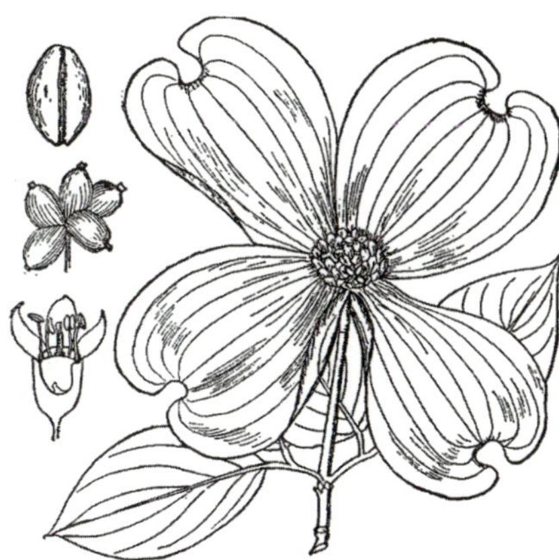

FLOWERING DOGWOOD *Cornus florida* L.

Flowering dogwood is an unusually beautiful shrub or small tree and occurs naturally only in York County. Planted specimens generally are only hardy in the southern and coastal areas of the state. Unfortunately flowering dogwood is under serious threat from Dogwood anthracnose *Discula destructive*, a non-native fungal disease.

Flowering dogwood reaches a height of 12–20 feet. The **bark** is gray and smooth on younger stems; on older trees it becomes black and finely blocky, as if broken into small squares. The **leaves** are opposite, entire, ovate to elliptic, bright green and smooth above, pale green with hairs on the veins beneath. They are 3–6 inches long.

The **flowers** are conspicuous and appear early in the spring. They are greenish-white or yellowish and are arranged in dense umbels surrounded by 4 large, white, petal-like bracts which give the appearance of large spreading flowers. The **fruit** is a bright red, ellipsoid drupe about ½ inch long that occurs in clusters.

The **twigs** are smooth, greenish and angular. The buds are covered by two valve-like scales. The flower buds are large and button-shaped.

The **wood** is hard and close-textured. In the past, it was widely used for the manufacture of shuttles for textile weaving. It is not used commercially in Maine due to its rarity.

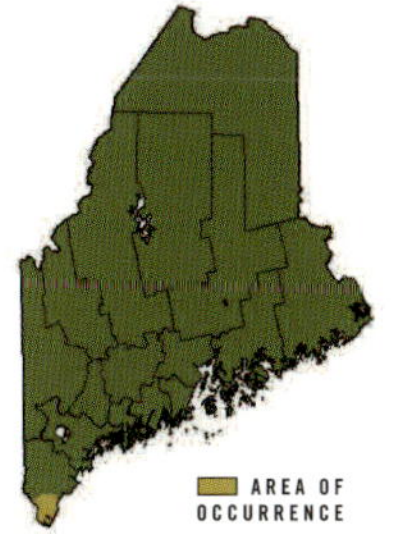

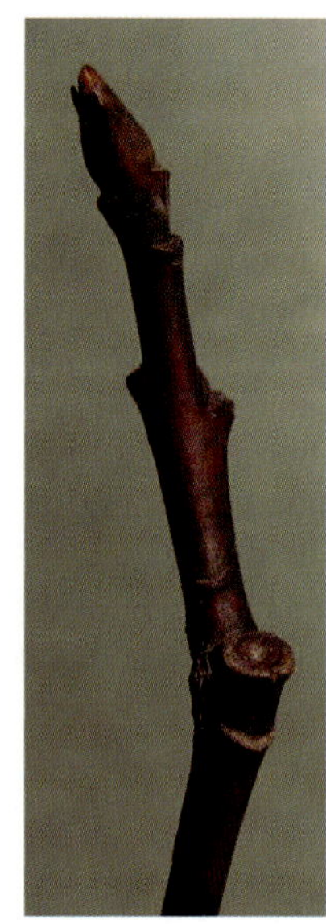

ALTERNATE-LEAF DOGWOOD
Cornus alternifolia L. f.

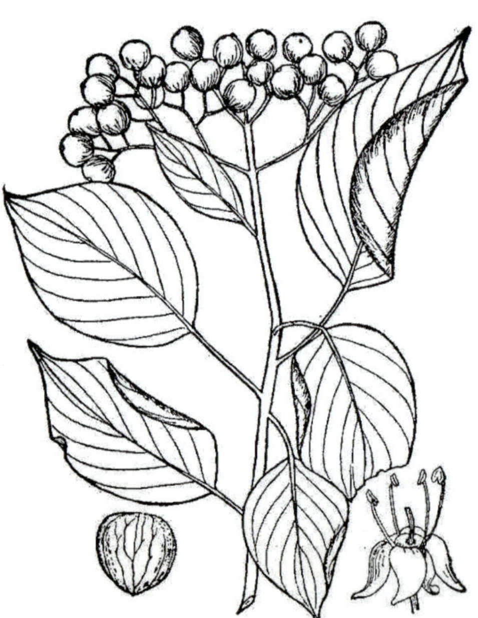

Alternate-leaf or blue dogwood occurs throughout the state as a shrub or small tree up to 20 feet tall.

The **leaves** are alternate, entire, elliptic-ovate and tend to be crowded at the ends of the twigs. They are 2½–4½ inches long, yellowish-green, smooth above and have appressed hairs beneath.

The creamy white **flower** clusters appear in June after the leaves have developed. The **fruit** is a bluish-black drupe, somewhat round, about ⅓ inch in diameter, that ripens in September and October.

The **twigs** are often lustrous and greenish-brown. Dead twigs become bright yellow-green.

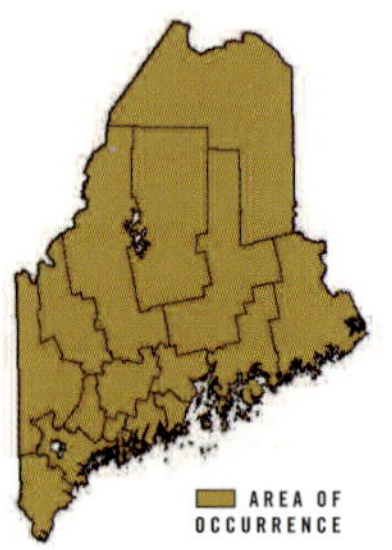

RED OSIER DOGWOOD *Cornus sericea* L.

Red osier dogwood is an abundant colonial shrub usually less than 10 feet tall. On very rare occasions it may reach the size of a small tree. It occurs throughout the state. It grows on the edges of fields and streams, and in wet areas. It readily invades fields, where it is considered a pest.

The **leaves** are opposite, entire, lance-shaped to elliptic to ovate, 2–4 inches long and whitened underneath, with 5–7 lateral veins.

The **flowers** are in flat-topped clusters. The **fruit** is white and ¼–⅓ inch in diameter. The **twigs** are bright red to green and minutely hairy. They are used for handles for cemetery baskets. The **buds** lack scales and are densely hairy.

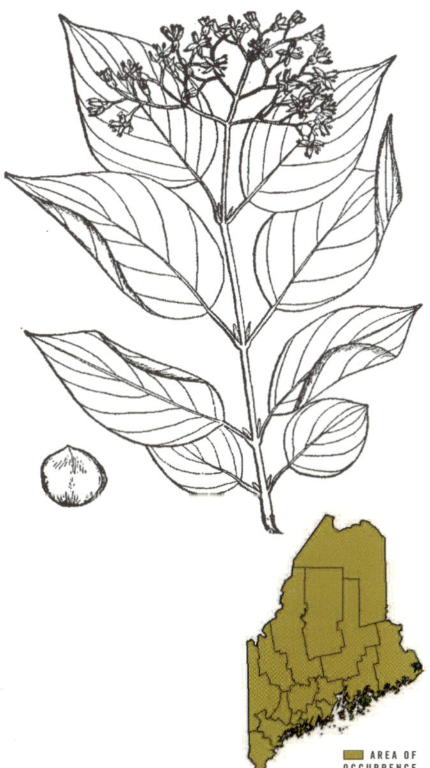

Red osier dogwood has opposite leaves and bright red twigs.

Selected References to Trees and Shrubs

General

Dwelley, Marilyn J. *Trees and Shrubs of New England,* Down East Books, printed by Twin City Printery, Lewiston, Maine, 1980.

Elias, Thomas S. *The Field Guide to North American Trees,* Van Nostrand Reinhold Co., New York, 1991.

Fernald, Merritt L. *Gray's Manual of Botany.* (Illus.) 8th ed. (1950) American Book Co., 1987.

Graves, Arthur H. *Illustrated Guide to Trees and Shrubs,* Harper & Row Publishers, New York, 1956. (Revised by Dover Publ.)

Harlow, William M., J. W. Hardin and F. M. White. *Textbook of Dendrology.* (Illus.) 7th ed. McGraw-Hill Book Co., Inc., New York, 1991.

Hyland, Fay. *Conifers of Maine,* (Illus.) U. of Maine Ext. Ser. Bull. 345. (Revised) 1961.

Hyland, Fay and Ferdinand H. Steinmetz. *Trees and Other Woody Plants of Maine,* Orono. University of Maine Press, 1978.

Little, Elbert. *The Audubon Society Field Guide to North American Trees: Eastern Region,* Alfred A. Knopf, New York, 1980.

McMahon, Janet S., G.L. Jacobson and Fay Hyland. *An Atlas of the Native Woody Plants of Maine; A Revision of the Hyland Maps.* Maine Agriculture Expt. Sta. Bul. 830. 1990.

Petrides, George A. *A Field Guide To Eastern Trees, The Peterson Field Guide Series:* Houghton Mifflin Co., Boston, 1988.

Symonds, George, W.D. *The Tree Identification Book,* William Morrow & Co., Inc, 1958.

USDA, NRCS. 2005. *The PLANTS Database,* Version 3.5 (http://plants.usda.gov). Data compiled from various sources by Mark W. Skinner. National Plant Data Center, Baton Rouge, LA 7087-4490 USA.

Specialized

Campbell, C.S. et al. *Revised Checklist of Vascular Plants of Maine.* Orono. Maine Agriculture and Forestry. Expt. Sta. Bul. 844. 1995.

Hyland, Fay and Barbara Hoisington. *The Woody Plants of Sphagnous Bogs of Northern New England and Adjacent Canada.* Illustrated by Laurel Smith. Orono, University of Maine Press, 1981.

Keys and Check Lists

Britton, N.L., and A. Brown. 1913. *Illustrated flora of the northern states and Canada.* (Source of some of the images).

Campbell, C.S., Fay Hyland, and M.L.F. Campbell. *Winter Keys to Woody Plants of Maine,* Orono. University of Maine Press. Revised Ed. 1978.

Dearborn, Richard G. *Key to the Woody Alpine and Subalpine Flora of Mount Katahdin.* Me. Field. Nat. 19 (6): 83–90, 1963.

Haines, A., and T.F. Vining. *Flora of Maine.* V.F. Thomas Co. Bar Harbor, Maine. 1998.

SPECIES INDEX

Maine's river drives continued until the 1970's.